$21.95

To Joel
With Love
from
Ashley Montagu
Princeton 1978—

THE REPRODUCTIVE DEVELOPMENT OF THE FEMALE

A Study in the Comparative Physiology of the Adolescent Organism

Third Edition

M. F. Ashley Montagu, Ph.D.

PSG Publishing Company, Inc.
Littleton, Massachusetts

Third Edition

Printed in the United States of America.

International Standard Book Number: 0-88416-218-4

Library of Congress Catalog Card Number: 78-55285

PREFACE TO THE THIRD EDITION

In the U.S. in 1977 about a million adolescents became pregnant, and more than 600,000 of these gave birth to babies. Physiologically, psychologically, socially, and economically these girls are usually quite unprepared either for pregnancy or the responsibilities of motherhood. These facts became increasingly clear to those of us who, over many years, have studied the reproductive development of the human female. My own studies were at first limited to the period of sterility in the female that followed the first menstruation, menarche. It was in the course of those studies that it became evident that pregnancy and parturition during adolescence are associated with a higher frequency of disadvantageous effects both for the mother and her offspring. In view of the importance of the subject a great deal more emphasis has, therefore, been placed on this aspect of the subject than was the case in earlier editions of this book. So that while much new material has been added which enables us to understand better both the neuroendocrinology and physiology of adolescent sterility in humans and in other animals, attention is now also focused in this book on the undesirability, from virtually every point of view of pregnancy in the adolescent. It is this finding that has now become the most important message that this book has to deliver.

The birth rate among girls between ages 15 and 17 nearly doubled in the decade ending in 1975, and this at a time when the rates for older women were dropping. We are confronted, then, with a very real problem: how to teach a sense of responsibility to our children. As John Stuart Mill wrote in 1859 in *On Liberty,* "It still remains unrecognised, that to bring a child into existence without a fair prospect of being able, not only to provide food for its body, but instruction and training for its mind, is a moral crime, both against the unfortunate offspring and against society." Could Mill have known then what we know today concerning the physiological and psychologically handicapping effects of adolescent pregnancy, he might have added that it is also a biological crime.

Perhaps if we go about it in the right way we may yet succeed in conveying the lesson enunciated in the words of another great Englishman, the historian Lord Acton, who remarked that "Liberty is not the power of doing what we like, but the right of being able to do what we ought."

My thanks are due to Dr. Rose E. Frisch, Center for Population Studies, Harvard University, and to Dr. Leona Zacharias, Vincent Memorial Hospital, Boston, for various kind services. To Louise Yorke of the Princeton Medical Center Library, to Helen Zimmerberg and Louise Schaeffer of the Biology Library, and Sally Burkman at the Population Research Library, all at Princeton University, I am grateful for much bibliographic help, and to Dr. Philip Gordon for his kindness in proofreading.

Princeton, New Jersey A. M.
27 November 1978

PREFACE TO THE SECOND EDITION

Since the publication of this volume in 1946 much new material has become available relating to the developmental physiology of the reproductive system in the adolescent organism, with especial reference to the human female. I have endeavored to incorporate all this material in the present edition.

What was once the *theory* of adolescent sterility is now an established *fact* of nature. I do not think there are any diehards left who question that fact. Before the publication of the first edition of this book there were many. Now that we no longer have a theory to establish, and the physiology of the reproductive system is so much better understood than it was twenty years ago, it were, perhaps, not altogether out of place to consider the social significance of the facts in the present book. I have attempted some discussion of this aspect of the subject in the final chapter.

The book, which was originally published under the imprint of C C Thomas of Springfield, Illinois, under the title *Adolescent Sterility*, has undergone a change of both publisher and title, and has been entirely rewritten.

I have to thank Professor I. B. Cohen of Harvard University for originally urging upon me the necessity of writing this book. Thanks are due also for help in supplying information to Drs. David M. Schneider, Ward Goodenough, Thomas Gladwin, B. J. Škerlj, S. S. Sarkar, H. Goot, Messrs L. E. Richdale, F. Rose, and Miss Tulika Sen.

25 September 1957. A. M.
Princeton, N. J.

PREFACE TO THE FIRST EDITION

This book presents the solution to a problem which greatly puzzled me when, as a student some twenty years ago, I first heard it propounded. It was the problem presented by the fact that anthropologists working among nonliterate peoples had observed that postpubertal females in such societies, while indulging in frequent sexual intercourse with mature males, rarely conceived until an appreciable period, frequently lasting several years, had elapsed after the attainment of "puberty." It had been fairly well established that neither contraceptive nor abortive practices could be held accountable for this remarkable infecundity. The question therefore arose as to precisely what factors might be responsible for it. In my student days there wasn't even an attempt at an answer to this question, at least not one that could be taken seriously. It was just another of those interesting questions one ran into from time to time to which there were no answers. I marked this down as one of the questions to which I would try to find the answer.

Before all else it was necessary to discover whether the problem was really a genuine one and not a spurious artifact of the investigator. I soon satisfied myself on that point, and proceeded to keep a weather-eye out for clues which might lead to a solution. The gynecological literature of the middle of the 19th century, especially that of France, gave me my first clue, but I had no experimental evidence with which to support the conclusions to which it led. Then, in 1930, I happened upon a paper by L. Mirskaia and F. A. E. Crew, "Maturity in the female mouse," (*Proceedings of the Royal Society of Edinburgh,* vol. 50, 1930, pp. 179–186). This paper, which is discussed at length in the pages that follow, provided me with the necessary experimental evidence. I had not long to wait before a more direct answer to my question, again based on experimental data, was returned in a short paper by C. G. Hartman, "On the relative sterility of the adolescent organism," (*Science,* vol. 74, 1931, pp. 226–227). From the appearance of these two papers I regarded the problem of the sterility of the adolescent organism as solved. All that would thereafter be required would be the filling in of the details.

Following several briefer studies in the preceding years,[1] in 1939 I published an account of the whole problem in the light of the detailed evidence, anthropological, physiological, and demographic, "Adolescent Sterility," in *The Quarterly Review of Biology,* vol. 14, 1939, pp. 13–34, 192–219. That study elicited the interest of anthropologists, endocrinologists, physiologists, physicians, psychologists, animal breeders, and students of human fertility and population. In the seven years which have elapsed since its publication important corroborative evidence, both from the

[1] M. F. Ashley Montagu, The adolescent sterility period in man, *American Journal of Physical Anthropology* (Supplement), vol. 21, 1936, p. 13; Infertility of the unmarried in primitive societies, *Oceania,* vol. 8, 1937, pp. 15–26; *Coming Into Being Among the Australian Aborigines,* London, Routledge, 1937; New York, E. P. Dutton and Co., 1938, pp. 236–258.

field and the laboratory, has become available. I have been repeatedly urged to incorporate this evidence together with the materials of an earlier study in a small volume, and thus render the essential facts accessible to a wider circle of readers. I am glad to have had the opportunity to be able to do this, and I wish here to thank my friends and correspondents for their stimulating and encouraging interest in the subject of this volume.

My thanks are due to William S. Laughlin for help in reading proof, and to Edward L. Cutler for assistance with the index.

20 August 1945 A. M.
The Biological Laboratories,
Cold Spring Harbor,
Long Island, N. Y.

CONTENTS

PREFACE TO THE THIRD EDITION *iv*
PREFACE TO THE SECOND EDITION *vi*
PREFACE TO THE FIRST EDITION *ix*

1 *Introduction 1*
2 *Adolescent Sterility in Other Animals 31*
3 *The Rhesus Monkey and the Java Macaque 53*
4 *The Chimpanzee 61*
5 *Adolescent Sterility in Humans 83*
6 *The Biology of Adolescent Sterility 161*
7 *The Psychological and Physiological Immaturity of the Young Mother 191*
8 *Summary and Conclusions 209*

APPENDIX A : *The Relation of Age at Menarche to Fertility 219*
APPENDIX B : *Adolescent Sterility in the Male 221*
APPENDIX C : *Age at Menarche and Duration of Adolescent Sterility 225*
APPENDIX D : *Advanced Age at Childbirth 229*

INDEX *231*

To Geoffrey M. Morant

1
INTRODUCTION

"Where Nature indicates a fact, we do well to attend to her instruction."
J. M. WADDY, 1846

The focus of attention of this volume is upon the infecundity of the adolescent female organism. Such a normal developmental phase of infecundity is usual, yet the existence of an adolescent sterile period only became apparent recently. Most persons are still unaware of adolescent sterility and fewer still are aware of its physiological significance and social and biological consequences. This alone would constitute sufficient reason for becoming acquainted with the facts.

There are other reasons which, in addition, make it desirable for every well-informed citizen to become familiar with the elementary facts concerning the physiology of reproductive development. When studied as part of the life cycle of human beings and of societies, they assume a significance far exceeding in importance the bounds of a seemingly narrow though perhaps interesting academic investigation.

In 1974 in the United States 608,000 births were registered for teenage girls, 12,529 to girls under 15, and 595,449 to girls aged 15 to 19. It is estimated that 300,000 abortions were performed on teenage girls in that year.[1] The highest rise in the birthrate in 1977 was to girls from age 15 to 17. The statement of these facts covers social and health problems of considerable magnitude.

Our society stands to gain much from an increasing sense of the importance of the citizen's health, mental and physical. This is especially important where the pregnant mother is concerned. Is there any age that is better than another during which a mother may conceive, maintain the pregnancy, and give birth to a healthy child? What is the best age for marriage for the human female? Should our society increase or reduce the statutory age of marriage? What is the relation between marriage at various ages and such conditions as maternal and infant mortality rates, abortions, miscarriages, toxemias, and the like.

In societies in which the female's age at marriage is low it is sometimes supposed that raising the age of marriage will reduce the birth rate. An accurate understanding of the facts of physiological development will show that a change of this kind is likely to have no significant effect upon the birthrate. Hence, the understanding of the facts of reproductive development in the human female is likely to provide a firmer basis upon which to rest an approach to the control of fertility than mere supposition concerning those facts.

In a declaration on discrimination against women adopted 7 November 1967 in the General Assembly of the United Nations, Article 6 Section 3 states, "Child marriage and betrothal of young girls before puberty shall be prohibited, and effective action, including legislation, shall be taken to specify a minimum age for marriage and to make the registration of marriages in an official registry compulsory."

This is an excellent provision, but clearly it will not

accomplish the desired ends unless the meaning of "puberty" is clearly understood and defined.

Some years ago, Dr. Marie Stopes, famous for her book *Married Love,* told a Royal Commission in London that more than one-fourth of British girls are fully mature and ready for marriage at 15 years. She stated that marriage should be possible for these girls though she would not encourage it. Her recommendation was to drop the minimum marriage age for girls from 16 to 15 years.[2]

As we shall see, such statements represent the expression of an inadequate understanding of the facts. It is something to be grateful for that Dr. Stopes would not encourage these "fully mature" girls to marry. There are, of course, some girls who are reproductively "mature" at 15 years and even earlier, but these are in the great minority. In any event, physiologic capacity should not constitute the sole measure of fitness for marriage, any more than it should be considered the sole measure for successful maternity—a function often confused both with reproductive ability and with marriage. To be a successful mother constitutes something in addition to being able to carry a fetus to term and nurse it. It means to be a psychologically satisfying mother to the child, and that is a state of being that has to develop in addition to reproductive capacity. Few females, however reproductively mature they may be at 15 years, would normally possess the capacity for successful motherhood in this psychological sense. A functionally less than optimally adequate reproductive system and the acquisition of a husband are no substitutes for this important human ability.

ADOLESCENCE

Adolescence (Latin *adolescentia, ad,* to + *olescere,* the inceptive form of *olere* [?], grow) means literally the process or condition of growing up. Specifically the term is applied to that period of life which falls between puberty and the full development of the body; a period generally assumed to extend in man from about 13 to 23 years of age, but which in its extreme limit may be 27 years of age.

While in common usage adolescence is usually taken to refer to the period from 13 to 21 years inclusive in both sexes, the facts indicate that in females adolescence extends from 12 to 21 years, and in males from 14 to 25 years. A period of great developmental importance as the adolescent period has attracted the attention of many students, as is evidenced by the very considerable literature devoted to its study.

The most significant contribution to our understanding of the biology of adolescence has come from the field of neuroendocrinology. The hypothalamus and several endocrine glands have been shown to represent the principal agents in the determination and regulation of the processes of development, particularly during the period of adolescence.

Puberty, menarche, adolescence, and maturity are terms that today are all too loosely used, a usage responsible for the perpetuation of many serious errors, yet the phenomena these terms embrace are capable of fairly precise definition. Such definitions will be offered in the course of this work.

As is well known, the onset of puberty is generally regarded as a condition which, in the female, is marked by the appearance of the first menstruation or the menarche (Gr. *mene*, month, *arche*, beginning); this event is practically universally assumed to be coincident with the establishment of the capacity to reproduce. The root *pu* of the word puberty (L. *pubertas* from *puber,* adult), actually means to *beget*. Thus, puberty, menarche, and the development of the capacity to reproduce are regarded as synchronous "events," which may be referred to by terms of synonymous and assumed equal value. Certainly every dictionary makes the identity of this triad complete, and most contemporary writers, with few exceptions, do so. Puberty is regarded as a moment in time determined by the arrival of the menarche, an event which is in turn taken to signalize the development of the capacity to reproduce.

Even textbook writers on the subject occasionally get derailed. For example, in one of the most authoritative works, *Marshall's Physiology of Reproduction* (1958), the relevant section commences with the inaccurate declaration "Puberty, or the period at which the organism becomes sexually mature."[3]

Similarly, Dublin in his *Factbook on Man*, published in 1965, equates "the beginning age of reproduction," with "age at menarche."[4]

Modern research on the reproductive physiology and growth of mammals such as the mouse, rat, guinea pig, hamster, sheep, cow, macaque, chimpanzee, and human, very strongly suggests that puberty, menarche, and the development or attainment of the capacity to reproduce are phenomena by no means as closely associated with one another as has hitherto been supposed, but rather that they are distinctly separable phenomena both in time and in character. It now appears fairly certain that puberty itself properly represents an extended process of development, rather than an event or a moment in time; and that it is during the progress of this developmental process that menarche is gradually established, and reproductive capacity only rarely. Puberty does not mark a definite step in a progressive movement, but represents more clearly an extended developmental period. Menarche represents a phase or aspect of that particular developmental period, the phase of beginning (anovulatory) ovarian activity. The capacity for reproduction generally follows appreciably later. The attainment of this capacity marks the termination of the period of puberty and the initiation of the period of nubility which, in the human species, terminates in the period of maturity only after a lapse of some years in time, and much increase in size and differentiation of various organs.

The interval between the establishment of menarche and the capacity to reproduce appears to be a normal feature of the developmental physiology of all mammals thus far investigated. In this work the question we are principally concerned with answering is: *What, if any, interval of time normally elapses between the appearance of the first menstruation and the ability to conceive and carry a fetus to term?*

The way for such a study was initiated by the work of C. G. Hartman[5,6] who, among modern students of reproductive physiology, was the first to suggest that the human adolescent female might be characterized by a period of relative sterility during the first part of this phase of her development. In the years that followed a good many students referred to this phenomenon with particular reference to its applicability to the human species, so the idea is not altogether unfamiliar. On the whole, however, the facts are not as well known as their import merits. In view of the value of a correct understanding of these facts for physicians, students of growth, physiologists, psychologists, anthropologists, educators, social workers, legislators and others this work has been written in order to make them more readily available.

THE FIRST OBSERVERS

The first observers to draw attention to the infertility of adolescent human females were ethnologists.

Working in the field among nonliterate people, ethnologists had more than once been puzzled by the fact that in the societies studied by them the unmarried postmenarcheal females, despite free and frequent intercourse with mature males, rarely became pregnant or bore children.

The first reference to this phenomenon appears to have been made by T. W. Leggatt in 1892 in his account of the people of Malekula in the Bismarck Archipelago. Leggatt noted that among the Malekulese, children were not born until two to four years after marriage. He thought this due "to a dislike among women to speedily becoming mothers."[7]

In this century the earliest investigator to draw attention to this phenomenon was C. G. Seligman. In a study of the Southern Massim of New Guinea, published in 1910, Seligman found that although the greatest freedom is allowed both sexes before marriage, children were rarely born of such unions. Seligman wrote: "Another puzzling feature of the licence undoubtedly permitted throughout the Massim area is the very small number of illegitimate births which take place. Wherever the confidence of the natives was gained it was admitted that abortion was induced, but the most careful inquiries failed to produce evidence that the practice was as frequent as might be expected considering the prevailing liberty. In fact with every à priori reason for expecting abortion to be commonly produced I came to the conclusion that in fact it was a somewhat infrequent event."[8]

Careful inquiries, Seligman adds, showed that no reliable method of contraception was used. Seligman could only conclude that New Guinea girls "were less easily impregnated than the women of the white race," an understandable conclusion during a period when anthropologists were still thinking in terms of racial differences rather than in terms of racial likenesses.

Eleven years later we find another field worker, J. H. Hutton, in his account of the Angami Nagas of Assam, describing the great sexual license allowed unmarried girls, stating that "illegitimate children are very rare," and that he was unable to find any evidence of abortion.[9]

In 1926 Rivers observed with respect to the Eddystone Islanders of Melanesia, studied by Hocart and himself, that "the very free relations existing before marriage might have been expected to lead to the birth of many children and to the existence of definite regulations for assigning such children to their proper place in society. Such births seemed, however, to be extremely rare, and in the whole of the pedigrees collected by us only one such case was given, and that many generations ago. We did not hear of any such birth either during our visit or in recent times; and so far as we know there was no one on this island who was the child of premarital intercourse. It was said that such births occurred, however, though no actual instances could be given."[10]

Rivers went on to say that "It is quite certain that births before marriage were very rare and two causes were given to account for this, abortion and a process resembling the other magico-religious rites of the island, called *egoro,* meaning 'barrenness,' which is believed to prevent conception."

Where information was available it appeared that women produced abortion by mechanical means or by the process of rubbing a certain leaf, which had previously been heated, over the belly, and finally by holding four leaves of another variety beneath the vulva, when, it was said, the child would "come out."

From Rivers' description of an *egoro* rite, the details of which need not be discussed here, it is perfectly clear that the rite is a purely magical one, and could have no possible physical effect upon the woman. This becomes even more clear when we learn that the rite is believed to be fully as efficacious in producing sterility in the woman when it is performed upon her husband alone. Rivers does to some extent recognize the magical nature of the rite, but he nonetheless considers it possible that the concoctions used may actually produce some pathological condition of the uterus in the woman. This, however, is pure assumption and as an explanation of the condition of infertility among the unmarried women will manifestly not do; for it is obvious that a fairly large proportion of the Eddystone Islanders must be fecund, and that the rarity of births among the unmarried is, at least in part, due to the fact that though there may have been much intercourse, relatively few pregnancies have resulted therefrom.

In 1929 Malinowski stated that among the Trobriand Islanders of North-Western Melanesia, "it is very remarkable to note that illegitimate children are rare. The girls seem to remain sterile throughout their period of licence, which begins when they are small children and continues until they marry; when they are married they conceive and breed, sometimes quite prolifically. . . . I was able to find roughly a dozen illegitimate children recorded genealogically in the Trobriands, or about one per cent. . . .

"Thus we are faced with the question: Why are there so few illegitimate children? On this subject I can only speak tentatively, and I feel that my information is perhaps not quite as full as it might have been, had I concentrated more attention upon it. One thing I can say with complete confidence: no preventive means of any description are known, nor the slightest idea of them entertained. This, of course, is quite natural. Since the procreative power of seminal fluid is not known, since it is considered not only innocuous but beneficient, there is no reason why the natives should interfere with its free arrival into the parts which it is meant to lubricate. Indeed, any suggestion of neo-Malthusian appliances makes them shudder or laugh according to their mood or temperament. They never practise *coitus interruptus,* and still less have any notion about chemical or mechanical preventives.

"But though I am quite certain on this point, I cannot speak with the same conviction about abortion, though probably it is not practised to any large extent. . . .

"So the problem remains. Can there be any physiological law which makes conception less likely when women begin their sexual life young, lead it indefatigably, and mix their lovers freely?"[11]

Doubts were raised concerning the accuracy of Malinowski's observations.[12] Doubts were, of course, to be expected, but it was soon established beyond all possible question that Malinowski's observations were in all particulars correct and confirmable.[13-15]

Fortune, in 1932, wrote of the Dobuans of the Western Pacific "It may be noted that in Dobu as in the Trobriands there is complete freedom for sex intercourse before marriage, and this freedom is used freely. Nevertheless, unmarried girls rarely bear bastards."[16]

Fortune stated that this could have been, among even less probable causes, due to the practice of abortion. From the account of the abortifacients he gave, it seems clear that such an explanation will not hold for the comparative infertility of the unmarried girls.

An illuminating account of conditions prevailing among the natives of Wogeo, one of the most northerly of the Schouten Islands in the territory of New Guinea, was given by Hogbin. Among the natives of this island sexual life did not begin until about the age of sixteen or seventeen years, or even later, after which it was particularly free. Hogbin wrote in 1935: "Single girls do sometimes have children, but illegitimacy is not nearly so common as one might have expected. Just why this is so it is impossible to say. Professor Malinowski, it will be remembered, found the same situation in the Trobriands. I observed one fact that bears directly on the problem, namely that it is extremely rare for women to have children until they are, I judge, more than 21 years of age, by which time most of them are safely married. I have noticed that even when a girl is married directly after her first menstruation, which does not regularly take place until almost certainly after the seventeenth year, it is most unusual for her to have a child for several years. One Dap girl had a child within about eighteen months after her first menstruation—fortunately for herself she was married—and this was so unusual that she was described as a coconut putting forth a shoot before it had fallen from the parent tree. The comments on this girl, in fact, brought the whole matter to my notice."[17]

"The natives," Hogbin added in a later work, ". . . have worked out the rule that at least four taro harvests—between two-and-a-half and three years—separate the coming-of-age ceremonies and pregnancy. . . .

"Young girls accordingly take no precautions against pregnancy: confident as yet of their inability to conceive."[18]

Although the ages that Hogbin gave are admittedly approximate, it was possible to check them fairly closely by reference to certain historic events known both to the investigator and to the islanders. The late appearance of menarche in some girls living in the tropics, and even its early appearance, has led some writers to attribute it to the heat. But the evidence is against this.[19-22] Girls, for example, in Moscow and in Tel Aviv attain menarche at the same age, namely, 13 years. The evidence indicates that age at menarche though certainly gene-influenced is also strongly influenced by nutritional and socioeconomic factors.

In monozygotic twins, for example, the mean difference in the age at menarche is between two and three months, whereas for dizygotic twins the mean difference is about 10 months, and for unrelated females it is about 19 months. These differences suggest multifactorial genetic influences, with the timing mechanism for the appearance of menarche following a regular and ordered sequence of events.

Shields found that the firstborn twin tends to attain menarche earlier by weeks or months than the later born twin. The earlier menarcheal twin is also usually the more intelligent.[23]

Valšík, reviewing the data on the relation of physiographic factors to onset of menarche, derived mainly from studies on Czechoslovakian, Polish, and other European populations, found that at low elevations menarche occurs more frequently in the summer months, while at higher elevations the higher frequencies occur in the winter months. Climate as such is of no great significance. It was also found that there is a tendency for girls of earlier menarcheal age to arrive at menarche in the first half of the year as compared with girls of slightly later menarcheal age who tend to arrive at menarche in the second half of the year.[24]

Eveleth has shown that girls of Japanese ancestry raised in Brazil achieve menarche at about the same median age as Brazilian girls of diverse ancestry. In Japan the average age at menarche is 14.03 years, whereas girls of Japanese ancestry in Brazil arrive at menarche at a median age of 12.85 years.[25] In American girls brought up in Rio de Janeiro age at menarche is 12.65 and in Brazilian white girls 12.68 years, in Brazilian mulattoes 12.99 years, and in Brazilian blacks 12.68 years.[26-28]

Wadsworth found that in 58 Asian and 27 European women living in Singapore the average age at menarche of both groups was about the same as it is in Europe, namely, 13.6 years. All these women were members of the "higher social levels."[29]

The importance of socioeconomic factors in influencing age at menarche has been underscored by the following studies:

Milicer found that in Breslau, Poland, the daughters of college-educated parents arrive at menarche some 6.7 months earlier than the daughters of unskilled manual laborers, the mean menarcheal ages being respectively 12.87 and 13.43 years.[30]

Also in Poland Laska-Mierzejewska, in an investigation of 9,589 rural girls, found that those from the more favored socioeconomic families achieved menarche earlier than those of less favored families. The difference between the extreme groups amounted to 0.51 years, with a mean of 14.04 years.[31]

In Bristol, England, Wolfinden and Smallwood found that menarche occurs three months earlier in grammar school girls than in secondary modern school girls, the latter attended by the less favored socioeconomic class.[32]

The daughters of professional and managerial classes arrive at menarche in Copenhagen some two months earlier than the daughters of unskilled workers.[33]

Carfagna et al. in a study of 8,261 Neapolitan Italian girls found that the median age at menarche was 12.49 years, with the higher class girls achieving menarche 0.76 years before the girls of lower class.[34]

Bourlière and his colleagues found that among urban females from the region around Paris from middle-class families of high intellectual caliber menarche was attained at an average age of 12.10 years—significantly earlier than among girls of lower socioeconomic status.[35]

Shakir, in a study of 4,673 girls of Arab descent attending schools in Baghdad, found that average age at menarche of those from the well-to-do class was 13.59 years as compared with 13.96 years for those from the slum areas.[36]

Lee, Chang, and Chen, reporting on 3,278 girls of Chinese origin living in Hong Kong found mean menarcheal age in the girls of high class to be 12.51 years as compared with 12.75 in the middle, and 13.25 years in the lower classes.[37]

Similarly among Singapore Chinese girls Aw and Tye found age at menarche in the highest income group to be 12.40 years, in the intermediate income group 12.66, and in the lowest income group 12.98 years.[38]

Not only do socioeconomic factors influence the time of appearance of menarche, they appear also to influence the onset of ovulation.

Banders-Van Halewijn and De Waard, in a comparative cytological survey of Netherlands Dutch and South African menstruating girls found a higher incidence of ovulatory menstrual cycles among Dutch girls (50 percent) as compared with Bantu girls (17.6 percent) of the same age group. Both menarche and ovulatory cycles occur earlier in European than in Bantu girls.[39] These differences almost certainly reflect a difference in nutritional history.

Reche reporting on the Matupi of New Pomerania, a Melanesian people, noted as early as 1910 that no girl who had not yet arrived at the age of seventeen had menstruated.[40-43] Here, too, we suspect the effects of a less than optimum nutritional history was expressing itself.

Höltker, in 1944, reported the age of menstruation in 32 girls drawn from various villages on the island of Manam, Melanesia, as being in the year 1938 (when his observations were made) 15⅔ years.[44]

From Hogbin's account it is clear that among the Dap of Wogeo despite frequent intercourse from menarche onward a girl does not normally conceive until several years have elapsed; not even when she marries early. Among this people there can be no objection to a recently married woman giving birth to a child, but this is so unusual a phenomenon that it has given rise to the belief that it is not a normal thing for a young adolescent married woman to be able to bear children, and certainly not normal in an unmarried woman. In this belief these natives show themselves far in advance of the western world, for their belief reflects the possession of knowledge based upon observation which, owing to a fortunate colligation of conditions which does not exist among ourselves, enabled them to arrive at a knowledge of facts that were not understood elsewhere in the world until recently.

Hogbin tells us that among the natives of Busama in North-Eastern New Guinea, "Single girls do not often become pregnant, though informants deny the practice of *coitus interruptus*.[45] Alleged abortifacients are known only to some older women and are never made known to the unmarried."

Writing in 1937 of the Eastern Rengmas of Assam, Mills observes "Before marriage . . . girls habitually admit lovers to their sleeping-platforms. . . . Curiously enough, no case seems to be known of a girl becoming pregnant through one of these intrigues. The Eastern Rengmas hold that even between adults one sexual act cannot cause pregnancy, while between the young so many would be required that there would be no time for them before marriage!"[46]

In 1938, writing of the Ifugao of the Philippines, Barton states that in the dormitories (*agamang*) of the young people, "conception results less frequently than one would expect. The surprising intensity of the sex life of the *agamang* period seems to create physiological conditions unfavorable to conception: possibly the excessive intercourse keeps the female organs in a state of hyperemia."[47]

Elwin in 1939 published an account of the Baiga tribe of the Central Provinces of India, in which he writes, "I have tried very hard to find some tradition of contraception among the Baiga, but with little success. This is, I think, a matter for some surprise, for on the face of it, it would seem inevitable that some such practice must exist. For how is it that this perpetual sexual intercourse, this almost uncontrolled licence, bears so little fruit? The illegitimate child is unknown, unwanted babies are few. Can it be simply that 'the grass that is constantly trodden doesn't grow'? Can it be that the sexual organs are in some way injured by their premature employment? The Baiga themselves explain it by saying that 'when a girl is going always to everyone, the mouth of her bag or womb never shuts, and so the man's seed runs in and out. That is why a loose girl never has a child.' "[48]

Ford, writing in 1940, states of the natives who live near the Trobrianders (as do the Dobuans referred to by Fortune) in the D'Entrecasteaux Islands that "It is everywhere considered disgraceful for an infant to be born to a single girl; but although sexual freedom exists among young unmarried people in some, though not in all, parts of the group [of islands], pregnancy as a result rarely eventuates, even though puberty has been established.

". . . though in the D'Entrecasteaux Islands married women sometimes attempt to prevent conception by drinking herbal potions, it was said that no such practice was undertaken by single girls. Informants, by whom the procreative process was well understood, rather accepted the fact that girls rarely conceived before marriage, even though menstruation was established, whereas they usually did so in the first few years of married life; but an occasional pregnancy in a single girl showed that this was not invariable.

"Since marriage usually occurs early, within at least a few years of puberty, the facts appear to support the view that, following the advent of puberty, there normally occurs a sterile period."[49]

In an account of the Abelam tribe of the Sepik district in New Guinea, Kaberry in 1941 states that "A girl generally becomes pregnant about three years after marriage, that is at the age of 16 or 17."[50]

Since girls in this region marry immediately after the first menstruation, and since this occurs in the fourteenth year, the indication here is that there is a mean adolescent sterility interval of about three or four years among Abelam females.

Mountford and Harvey, writing in 1942, reporting their investigations carried out in 1938 and 1939 among the Adnjamatana tribe of the Northern Flinders Ranges of South Australia, state that "Marriage takes place at puberty. But enquiries as to the approximate age of the mother at birth of her first child revealed the fact that several years normally elapsed before this event took place. The old women of the tribe were quite emphatic and positive on this point."[51]

Abortive or contraceptive practices of any kind are quite unknown among the members of this tribe.

Mills, writing in 1942, of conditions "in a large naval port of a tropical country" states that "Although there was an almost unbelievable freedom of sex relations prevailing among the city's lower economic groups, I found a late age for the first childbirth. One young unmarried mother had been consorting for three years with a foreigner before giving birth to her first child, another had been consorting with two native men and one foreigner; and so the stories went."[22]

In the same article Mills reports that "During a visit among the natives of Luzon, Philippines, I learned that relations between the two sexes were entirely uncurbed through adolescence and youth. A keen, American-trained native Bontoc physician gave me a good description of such affairs among his people. The parents keep their youngest child in the main hut with them, but all older children sleep out in a so-called dormitory. . . . Here children and youths from the different families may mix as freely as they wish without any parental supervision whatever. When the form of a girl fills out nearing the physical development that indicates to her people the close approach of reproductive capacity, she is properly married off. Illegitimacy is rare and is much frowned upon, even though adolescent sex relations receive no attention whatever."[22]

Ronald and Catherine Berndt, writing in 1943 of their work among the Australian aborigines of the Ooldea region, in Western South Australia, state that "couples indulge in early coitus, from the period when the male (at about the age of 14 years) and the female (at about the age of 11 years) reach sexual maturity. During the following period the unmarried girl, despite frequent intercourse, fails to conceive or bear offspring. It is our opinion that a period of infertility does exist for some girls in the Ooldea camp, but not for all. As most children are betrothed at an early age, they are married as soon as both pass through the initiatory stages. Obviously there is a period of two or three years when the pair are sexually mature and cannot marry. They indulge in prenuptial coitus between themselves or with others. . . . At Ooldea there were no cases recorded of full-blood children being born out of wedlock."[52]

In an account of their observations on the aborigines of Yirrkalla in North-Eastern Arnhem Land in the Northern Territory of Australia during 1946–47, the Berndts write, "The problem of adolescent sterility is a very interesting one; and particularly in this area, where girls usually go to their husbands before puberty. Usually there are no offspring for (apparently) a couple of years after menstruation takes place, although at present there is one girl in the camp who became pregnant a few months after this event. The other women are much concerned, alleging that she is too young and will probably die: they blame the husband for paying her too much attention, alleging that he neglected his older wives on her account and would not leave her alone."*

McLaren, who from 1911 to 1919 lived among the Australian aborigines at Simpson's Bay, Cape York, in an

*Ronald and Catherine Berndt, 23 January 1947, personal communication.

extraordinarily interesting book recounting his experiences, quite in passing makes the following remarks: "Freely and without concealment the unmarried girls give themselves to the young men of the tribe and to visitors from other tribes, sometimes in exchange for a coveted article of ornament or use, but oftener because such acts were to them as natural as eating or sleeping.

". . . But because of their knowledge of the abortive qualities of herbs and the like, illegitimate children were extremely rare. In all my eight years at Cape York they happened only twice, and in each case the girls were saved from being speared to death by the fact that their babies were born dead and undeveloped."[53]

It is more than doubtful that the Cape York aborigines "knowledge of the abortive qualities of herbs and the like" had anything whatever to do with their infertility.

In 1965, M. J. Meggitt reported that among the Central Australian Walbiri aborigines menarche usually occurs when the girl is about 14 years old. However, "Although girls start copulating several years before they reach puberty, I have no record of any woman who bore a child before attaining the (apparent) age of about 17."[54]

Summarizing his observations on many different Australian tribes, A. A. Abbie writes, "Girls are usually betrothed at birth, sometimes—on a speculative basis—before birth. Normally they go to live with their husbands at the age of ten to twelve years but intercourse does not usually begin until after puberty. As in other peoples, there is normally a period of relative adolescent sterility and most first babies are born after their mothers have reached the age of sixteen or more years; in one group studied, the average age at the birth of the first child was 18.5 years, but exceptions can, of course, occur in any people."[55]

Concerning Australia, it is of interest to record here that the Rev. Schurmann, who from 1840 had been Protector of Aborigines, and later a missionary among the natives of the Port Lincoln district, stated in a work published in

1861 that "Although married so very young, the women . . . generally have no children before the age at which they get them in Europe."[56]

Evidently the fact that Australian aboriginal females were generally infertile for some years following "puberty" had not gone altogether unnoticed before the modern rediscovery of that fact.

Harrisson, writing in 1937 about the natives of the New Hebrides notes of the North-West Malekulese that "Very rarely a child is born to an unmarried woman. Girls are married young."[57]

In 1947 Elwin reported the existence of a period of adolescent sterility among the Muria of the Central Provinces of India (see pp. 130–135 of the present volume).[58]

Among the Fijians of the island of Vanua Levu Quain evidently observed "a period of sterility which might seem to follow first menstruation at Nakoroka."[59] He suggests, however, that this might be due to the practice of having a twig inserted into the cervix uteri in order to draw blood should the normal flow not reappear. It is difficult to see the relevance of this suggestion. The beliefs of these Nakorokan villagers that pregnancy requires long-continued regular intercourse[59] would strongly suggest that a considerable interval of time usually elapses between menarche and conception.

Among the Gonds, the largest aboriginal tribe of India, Dube found that girls generally marry after puberty, and that there is generally a sterile period of about two years following marriage.[60]

In a personal communication Dr. Thomas Gladwin has reported to me that on the island of Romonum (Truk), in Micronesia, average age at menarche is 14 years. Intercourse commences at that time, and is fairly consistent with several partners thereafter, even following marriage. Age at marriage is 16 to 18 years, age at conception is 19 to 20 years. It is difficult to obtain reliable estimates of age at Romonum, and Dr. Gladwin describes his own estimates as "educated guesses." Nevertheless, the indications suggest a possible average adolescent sterility interval of about 3 years exists in the females of Romonum.

No evidences of contraceptive practices were discoverable, and while abortion by massage is known, there was no evidence of its current use.*

Among the Kaska Indians of Northern British Columbia and southern Yukon Territory, Canada, Honigmann (1954) found that "Although most girls commenced sex relations at about 17, a girl rarely bore a child until about two years later. In the words of the informant, 'Many people could get away with it for two years.' "[61]

In 1956 Danielsson published his observations on the natives of Raroia, Polynesia. Here boys and girls begin sexual intercourse long before puberty and continue an active sexual life with many different lovers for years after puberty. Danielsson writes, "Polynesian girls seldom become pregnant despite their free life. The natives themselves have a convenient explanation of this. They say that girls become pregnant only if they are keeping regular company, and the mothers therefore advise their daughters to change their boy friends as often as possible. This, however, is obviously a defence of prevailing conditions instead of a cause for them—however attractive the explanation may sound."[62]

Danielsson having made himself acquainted with the physiological facts as originally set out in the first edition of the present book proceeds to give the correct explanation of the relative infertility of these Polynesian girls.

Among the Apa Tanis of the Eastern Himalayas, von Fürer-Haimendorf found that sexual play and intercourse begin among boys and girls at an early age, but that premarital pregnancies were rare. "The suggestion that a girl might become pregnant after one or two nights' dalliance with a man was greeted with hilarity. My Apa Tani friends thought it impossible; they denied all knowledge of contraceptive practices but explained that a continuous re-

*Thomas Gladwin, 27 February 1950, personal communication.

lationship was necessary before a child could be conceived."[63]

Bates, in his account of the Ifaluk people of the Carolines, remarks upon "the puzzle of the absence of illegitimate births among the young unmarried."[64]

Geddes reports of the Land Dayaks of Sarawak, where boys and girls commence sexual relations quite early, "Whether because of youngness or because the affairs are short and spasmodic, young girls do not often become pregnant."[65*]

Among the Pygmies of the Ituri Forest in Central Africa when a girl first menstruates the event is celebrated by her taking a group of younger and older girls with her to the house of the "elima." Here she is taught the arts of motherhood, and how to live like an adult. Here, too, a lover may gain access to her. If he has intercourse with her he may not leave the house until the festival is over. Turnbull states that with such intercourse go certain restrictions. "These restrictions," he adds, "must be effective in preventing conception, because nowhere have I ever heard of a girl becoming pregnant in an elima house."[66]

Describing the Eskimos of Baffin Island in 1962 Miss Mary Jean Kempner writes, "Sexual life among young people is free and realistic. . . . Considerable experimentation is, therefore, encouraged among young teenagers with the community's hearty approval. . . . Strangely enough, in spite of premarital promiscuity, pregnancy is rare even through the first year of marriage. Experts in the Arctic have difficulty explaining this phenomenon, although some suspect the existence of a contraceptive moss or seaweed."[67]

Among the Nunamiut Eskimos of Anaktuvuk Pass, Alaska, sexual promiscuity is frowned upon, however, during her late teens a girl is expected to have an affair or two as a prelude to marriage, but there appear to be no records of such intercourse leading to pregnancy.[68]

---

*See also Hendrik de Leeuw, *Crossroads of the Java Sea*. New York: Jonathan Cape & Harrison Smith, 1931, pp. 47–48.

Heider, in writing of the Dugum Dani, a Papuan people of the highlands of West New Guinea, states that his informants insisted that premarital intercourse was very common, but that girls did not have babies because "conception required a great deal of sexual intercourse and these premarital affairs were too brief for that."[69]

The Jivaro of Ecuadoran Amazon "are aware of the relationship between insemination and pregnancy, and predict that a fully mature woman can give birth to a baby when the *naitka* tree . . . has flowered three times after her marriage. The *naitka* flowers twice a year, so their estimate is a year and a half. If the bride is just barely past puberty, they estimate that a longer period, five *naitka,* or two and a half years is required.

"Since premarital sexual relations are common, the birth of a child in a shorter period after marriage is not a cause for surprise. After the birth of the first baby, a three *naitka* minimum is estimated for the next."[70]

Richard B. Lee notes that among the ¡Kung Bushmen of the Dobe area of Botswana menarche occurs between 15 and 17 years of age, and that first pregnancies are delayed by several years. "Thus a woman does not bear her first child until she is between eighteen and twenty-two years of age."[71]

In a more detailed study of 165 ¡Kung women Nancy Howell found menarche to occur at a mean age of 16.5 years and the first live birth to occur around the age of 19.5 years. About half of the women were married before menarche and half after. The women who menstruate early are those who tend to marry after menarche.[72]

Robert Levy writes of his experience among the Tahitians that "Although most girls in rural areas are said to begin having intercourse by the time they are sixteen, there appear to be relatively few pregnancies among girls of this age."[73]

An anonymous writer on Tahitian girls writes, "Surprisingly few Tahitian girls become pregnant before they reach 18."[74]

Talwar, in a study of over 2,000 parous Bengalee women ranging in age from nine to 22 years of age found that the percentage of parous nonfecund women diminished from 98 to nine and zero, as the age of these women increased from 11 to 18 and 19 years, thus indicating the existence of a period of adolescent sterility.[75]

THE PROBLEM

The reports cited above, as well as those given in subsequent pages of this book, raise a number of questions. What can be the explanation of this apparent infertility of the adolescent female in these societies? Surreptitious abortion perhaps, which escapes the inquiring investigator's attention? This is possible, but there is no evidence that any infertile girls of these societies ever resorted to such a practice.

Contraception? Clearly not among the Australian aborigines and the Trobriand Islanders, who deny the relationship between intercourse and conception.[11,76] And what of the abortive properties of the herbs used by the aborigines described by McLaren, and the contraceptive properties of those used by the Eddystone Islanders described by Rivers? The evidence has been carefully examined by Himes, and it is his and the now generally considered opinion that none of the herbs thus far investigated for such properties can possibly produce the effects that have been attributed to them.[77]

Early and indefatigable sexual life? There is no evidence that early and indefatigable sexual life produces a condition of adolescent sterility. Such suggestions represent nothing but guesses. On the other hand, the evidence available for those simpler peoples among whom sexual life commences at a very early age shows that they are neither more nor less infecund than those peoples among whom sexual life is begun late. It may be recalled here that the natives of Wogeo, described by Hogbin, did not commence active sexual life until they were seventeen years of age, yet they remained sterile for several years thereafter. It would, indeed, be very surprising to discover both licentiousness and abstinence producing the same effect.

Free mixing of lovers? It has been suggested that prostitutes are sterile because they mix their lovers frequently. Not only have Western students suggested this but the Azande of the Sudan, according to Evans-Pritchard, believe that the birthrate has fallen because adulterous intercourse with a large number of men renders a woman less fruitful because "the womb suffers from so many organs."[78] It has been thought that such mixing of lovers results in immunization against sperm. Sutherland and Landing have investigated this matter in 28 professional prostitutes and 21 controls. They could find no evidence of antibody production against sperm.[79]

The fertility of prostitutes who rarely use contraceptives is undoubtedly low, but this is, at least in part due to the occurrence of inflammatory disease following gonorrhea or postabortion. Mechanical factors due to tubal and ovarian adhesions, as well as emotional factors, no doubt also play a role.

There is some evidence that prostitutes develop a much higher opsonic activity in their sera than nonsexually active women (nuns) and young adults of both sexes. Bisno and his co-workers have found that the antibodies produce conditions which render the bacterium of gonorrhea, *Neisseria gonorrhea,* susceptible to phagocytosis.[80] It is conceivable that a similar mechanism may play some part in producing the relative infertility of prostitutes. However, as we shall see, the facts of normal physiological development are quite sufficient to explain the relative infertility of adolescent females in all human societies.

Elwin has conclusively shown that among the Muria of Bastar, Central Provinces, India, the girls who mixed their lovers freely in fact had more pregnancies than girls who had but a single steady lover.[62]

What other explanation, then, can there be for the infertility of these promiscuous adolescent females? Is the adolescent infertility of these females due to a natural or normal state of relative sterility characteristic of all normal human females during this period of development, and perhaps also of all physiologically adolescent animals? Or is the phenomenon merely limited to the adolescent females of nonliterate societies? Finally, is it not possible that the phenomenon is a purely specious one? To provide a sound answer to these questions is the purpose of this volume.

DEFINITIONS

A number of terms which will frequently be used in the following pages may be conveniently defined here according to the definitions adopted by the Population Association of America.[81]

Fertility (also *Natality, Fecundity*): The physiological, measurable capacity to reproduce.

Sterility (also *Infertility, Infecundity*): The physiological incapacity to reproduce; synonymous with childlessness.

Reproduction: The net effect of procreative tendency.

REFERENCES

1. Baldwin, W. H. Adolescent pregnancy and childbearing—growing concerns for Americans. *Population Bulletin.* 31: 2, 1977.

2. *Time*. New York, 1 Dec. 1952, p. 61.

3. Marshall, John, and F. H. A. *Marshall's Physiology of Reproduction*. Edited by A. S. Parkes. In *The Life Cycle*. 3d edition. London: Longmans, 1958, 2: 822.

4. Dublin, L. I. *Factbook on Man*. New York: Macmillan & Co., 1965, p. 16.

5. Hartman, C. G. On the relative sterility of the adolescent organism. *Science*. 74: 226–7, 1931.

6. Hartman, C. G. Studies in the reproduction of the monkey *Macacus (Pithecus) rhesus*, with special reference to menstruation and pregnancy. *Carnegie Institution of Washington, Contributions to Embryology*. 134: 1–161, 1932.

7. Leggatt, T. W. Malekula. *Proceedings of the Australasian Association for the Advancement of Science*. 1892, p. 704.

8. Seligman, C. G. *The Melanesians of British New Guinea*. Cambridge and New York: Cambridge University Press, 1910, p. 500.

9. Hutton, J. H. *The Angami Nagas*. London: Macmillan & Co., Ltd. 1921, p. 171.

10. Rivers, W. H. R. *Psychology and Ethnology*. New York: Harcourt, Brace & Co., 1926, pp. 76–8.

11. Malinowski, B. *The Sexual Life of Savages*. New York: Harcourt, Brace & Co., 1929, pp. 166–8.

12. Rentoul, A. C. Physiological paternity among the Trobrianders. *Man*. 31: 152–4, 1931.

13. Austen, L. Procreation among the Trobriand Islanders. *Oceania*. 5: 102–13, 1934.

14. Malinowski, B. Pigs, Papuans, and police court perspective. *Man*. 32: 33–8, 1932.

15. Rentoul, A. C. Papuans, professors, and platitudes. *Man*. 32: 274–6, 1932.

16. Fortune, Reo F. *Sorcerers of Dobu*. New York: E. P. Dutton & Co., 1932, pp. 239–40.

17. Hogbin, I. H. Native culture of Wogeo: report of field work in New Guinea. *Oceania*. 5: 310–37, 1935.

18. Hogbin, I. H. Puberty to marriage: a study of the sexual life of the natives of Wogeo, New Guinea. *Oceania*. 16: 185–209, 1946.

19. Robertson, J. On the period of puberty in the Negro. *Edinburgh Medical and Surgical Journal*. 69: 69–77, 1848.

20. Curjel, D. F. The reproductive life of Indian women. *Indian Journal of Medical Research*. 8: 366–71, 1920.

21. Mills, C. A. Geographic and time variations in body growth and age at menarche. *Human Biology*. 9: 43–56, 1937.

22. Mills, C. A. They mature *later* in the tropics. *Harpers Magazine*. 184: 294–8, 1942.

23. Shields, J. *Monozygotic Twins*. New York: Oxford University Press, 1962, pp. 115–16.

24. Valsik, J. A. The seasonal rhythm of menarche: a review. *Hum Biol*. 37: 75–90, 1965.

25. Eveleth, P. B. The effects of climate on growth. *NY Acad Sci*. 134: 750–9, 1966.

26. Eveleth, P. B. Eruption of permanent dentition and menarche of American children living in the tropics. *Hum Biol*. 38: 60–7, 1966.

27. Eveleth, P. B., and Freitas, J. A. de So. Tooth eruption and menarche of Brazilian-born children of Japanese ancestry. *Hum Biol*. 41: 176–84, 1969.

28. Gurjao, I. Contribucao ao estudo do aparecimento das

primeiros regras em mulheres brasileiras. Presented at Brazilian Anthropological Association. Curitiba, 1959.

29. Wadsworth, G. R. Menstrual characteristics of women living in a tropical climate. *J Obs Gyn Br Empire.* 60: 110–13, 1953.

30. Milicer, H. Age at menarche of girls in Wroclaw, Poland, in 1966. *Hum Biol.* 40: 248–9, 1968.

31. Laska-Mierzejewska. Effect of ecological and socio-economic factors on the age at menarche, body height and weight of rural girls in Poland. *Hum Biol.* 42: 284–92, 1970.

32. Wolfinden, R. C., and Smallwood, A. L. Annual Report of the Principal School Medical Officer to the City and County of Bristol Education Committee, 1958.

33. Tanner, J. M. *Growth at Adolescence,* 2nd edition. Oxford: Blackwell Scientific Publications, Ltd, 1961.

34. Carfagna, M., Figurelli, E., Materese, G. et al. Menarcheal age of schoolgirls in the district of Naples, Italy, in 1969–70. *Hum Biol.* 44: 117–25, 1972.

35. Bourlière, F. et al. 'Normes' de viellissement morphologique et physiologique d'une population de niveau socio-economique eleve de la region Parisienne. *Bulletin et Mémoires de la Société d'Anthropologie de Paris.* 10: 11–39, 1966.

36. Lee, M. C., Chang, K. S. F., and Chan, M. M. C. Sexual maturation of Chinese girls in Hong Kong. *Pediatrics.* 32: 389–98, 1963.

37. Shaki, A. The age of menarche in girls attending schools in Baghdad. *Hum Biol.* 43: 265–70, 1971.

38. Aw, E., and Tye, C. Y. Age of menarche of a group of Singapore girls. *Hum Biol.* 42: 329–36, 1970.

39. Banders-Van Halewijn E. A., and De Waard, F. Menstrual cycles shortly after menarche in European and Bantu girls. *Hum Biol.* 40: 314–30, 1968.

40. Reche, O. Untersuchungen über Wachstum und Geschlechtsreife bei Melanesischen Kindern. *Korrespondenzblatt der Deutschen Gesellschaft für Anthropologie, Ethnologie, und Urgeschichte.* 41: 48–55, 1910.

41. Neuhaus, R. *Deutsch-Neu-Guinea.* Berlin: 1: 149–75, 1911.

42. Fehlinger, H. *Sexual Life of Primitive People.* London: A. & C. Black, 1921.

43. Mills, C. A. Physiologic sterility of adolescence. *Hum Biol.* 8: 607–15, 1936.

44. Höltker, G. Etwas über Reifalter bei melanesischen Mädchen und die Fruchtbarkeit melanesischer Frauen in Neuguinea. *Bulletin de la Société fribourgeoise des Sciences naturelles.* 37: 1–19, 1944.

45. Hogbin, I. H. Sex and marriage in Busama, North-Eastern New Guinea. *Oceania*. 17: 119–38, 1946.

46. Mills, J. P. *The Rengma Nagas*. London: Macmillan & Co., Ltd., 1937, p. 43.

47. Barton, R. F. *Philippine Pagans*. Hollywood, Florida: Transatlantic Arts, Inc., 1938, p. 11.

48. Elwin, V. *The Baiga*. London: John Murray, Ltd., 1939, pp. 219–20.

49. Ford, E. Notes on pregnancy and parturition in the D'Entrecasteaux Islands. *Med J of Australia*. 16 November 1940.

50. Kaberry, P. M. The Abelam tribe, Sepik District, New Guinea. *Oceania*. 11: 233–58, 1941.

51. Mountford, C. P., and Harvey, A. Women of the Adnjamatana tribe of Northern Flinders Ranges, South Australia. *Oceania*. 12: 156–62, 1942.

52. Berndt, R. and C. A preliminary report of field work in the Ooldea region, Western South Australia. *Oceania*. 13: 243–80, 1943.

53. McLaren, J. *My Crowded Solitude*. London: George Newnes, 1926, 74.

54. Meggitt, M. J. *Desert People: A Study of the Walbiri Aborigines of Central Australia*. Chicago: University of Chicago Press, 1965, p. 271.

55. Abbie, A. A. *The Original Australians*. London: Frederick Muller, Ltd, 1969, 51.

56. Wilhelmi, C. Manners and customs of the Australian natives, in particular of the Port Lincoln district. *Trans R Soc Victoria*. 5: 180, 1861.

57. Harrisson, T. *Savage Civilization*. New York: Alfred A. Knopf, Inc., 1937, p. 38.

58. Elwin, V. *The Muria and Their Ghotul*. New York: Oxford University Press, 1947, p. 463.

59. Quain, B. *Fijian Village*. Chicago: University of Chicago Press, 1948, p. 316.

60. Dube, L. Pregnancy and childbirth among the Amat Gonds. *Eastern Anthropologist*. 2: 153–9, 1949.

61. Honigmann, J. J. The Kaska Indians: an ethnographic reconstruction. *Yale University Publications in Anthropology*. 51: 128, 1954.

62. Danielsson, B. *Love in the South Seas*. New York: Reynal & Co., 1956, p. 103.

63. Furer-Heimendorf, C. von. *Himalayan Barbary*. New York: Abelard-Schuman, Inc., 1956, 67.

64. Bates, M., and Abbott, D. *Coral Island*. New York: Scribner & Sons, 1958, p. 118.

65. Geddes, W. R. *Nine Dayak Nights*. New York: Oxford University Press, 1961, p. 70.

66. Turnbull, Colin. *The Forest People*. New York: Simon & Schuster, 1961, p. 198.

67. Kempner, M. J. Notes from Baffin Land. *Harpers Magazine*. 225: 59–64, 1962.

68. Gubser, N. J. *The Nunamiut Eskimos: Hunters of Caribou*. New Haven: Yale University Press, 1965, 113.

69. Heider, Karl G. *The Dugum Dani*. Chicago: Aldine Publishing Co., 1970, pp. 75, 224.

70. Harner, M. J. *The Jivaro*. New York: Doubleday/Natural History Press, 1972, p. 83.

71. Lee, Richard B. Population growth and the beginnings of sedentary life among the ¡Kung Bushmen. Edited by Brian Spooner. In *Population Growth: Anthropological Implications*. Cambridge, Mass.: MIT Press, 1972, pp. 331–2.

72. Howell, Nancy. The population of the Dobe area ¡Kung. Edited by R. B. Lee, and I. DeVore. In *Kalahari-Hunter-Gatherers*. Cambridge, Mass.: Harvard University Press, 1976, 145–6.

73. Levy, Robert I. *Tahitians*. Chicago: University of Chicago Press, 1973, p. 143.

74. Anon. The Girls of Tahiti. *Playboy*. 13: 296, 1966.

75. Talwar, P. P. Adolescent sterility in an Indian population. *Hum Biol*. 37: 256–61, 1965.

76. Montagu, Ashley. *Coming Into Being Among the Australian Aborigines*. 2nd edition. London and Boston: Routledge, 1974.

77. Himes, N. E. *Medical History of Contraception*. Baltimore, Md.: The Williams & Wilkins Co., 1936, pp. 3–56.

78. Evans-Pritchard, E. E. *Essays in Social Anthropology*. New York: The Free Press, 1963, p. 120.

79. Sutherland, J. M., and Landing, B. H. Failure to demonstrate antibody to sperm in serum of prostitutes. *Lancet*. 2: 56, 1 July 1961.

80. Bisno, A. L., Ofek, I., Beachey, E. H. et al. Prostitutes develop opsonic antibody against *N gonorrhoeae*. *JAMA*. 228: 1215, 1974.

81. American Population Association. Organization for research in world population. *Hum Biol*. 6: 223–39, 1934.

2
ADOLESCENT STERILITY IN OTHER ANIMALS

THE PENGUIN

Up to the year 1949 observations relating to the phenomenon of adolescent sterility in birds were not available. In September 1949 that lacuna was ably filled by Richdale's monograph on the Yellow-eyed penguin (*Megadyptes antipodes*).[1] Richdale studied altogether 460 birds under natural conditions. The exact age of 162, and the age within three weeks of 298 birds, was known.

Richdale found that of two-year-old penguins nearly 48 percent of females did not produce eggs, though there was strong circumstantial evidence that coition was effected. Furthermore, birds younger than two years frequently laid but one egg, while of 40 two-year-olds 32 percent laid only one egg to a clutch; whereas of 218 older penguins only three percent laid one egg to a clutch, the usual number being two. Nearly 7 percent of two-year-old females deserted their eggs. Two-year-old males also did so. Finally, only one-third of the eggs laid by two-year-olds hatched. Richdale writes, "of 50 eggs laid by two-year-old females mated to older males, and fully incubated, only 34 percent hatched. The birds of this group which were found subsequently as three-year-olds, showed an improved hatching rate of 83 percent although the hatching rate of other three-year-olds, which did not lay as two-year-olds, was only 67 percent of eggs fully incubated."

Of the 34 percent of fully incubated eggs of two-year-olds only 25 percent hatched. If the figures for birds that did not lay and birds that did not hatch their eggs are combined, the two-year-old penguins are fertile only to the extent of 18 percent. The evidence indicates that if any male penguins are fertile at this age their number is negligible.

From Table 1 it will be seen that there is an increase in fertility after the second year, but that it is only after the third year, that is, in the fourth year, that maximum fertility is achieved. But since Richdale does not give exact figures of the number of birds involved in each category precise conclusions cannot be drawn, though he writes that from the fourth year on "there does not seem to be any significant decrease in the number of eggs laid." Richdale adds, "There is, however, a suspicion formed in the field, but concerning which I have not yet sufficient statistical evidence, that aged Yellow-eyed Penguins may tend to produce single eggs to the clutch. The three birds which laid a single egg each . . . were known to be in at least their ninth, twelfth, and fourteenth year, respectively."

It may be safely predicted that further research will confirm Richdale's suspicion that older birds do tend to be less fertile than younger ones.

Richdale writes, "In summary, it would seem that in the Yellow-eyed Penguin, and possibly in some other species of birds, there is, on the average, a tendency towards sterility with young birds in the period when a small proportion of the individuals are able to breed." He then cites the facts which have already been resumed and concludes, "These facts resemble closely what has been

Table 1
Influence of Age on Fertility in the Yellow-eyed Penguin

	Eggs Laid By							
Number of Eggs	2-year-olds	3-year-olds*	3-year-olds†	4-year-olds	5-year-olds	6-year-olds	7 to 11 years old	Old Birds
Laid	67	61	43	78	56	31	37	123
Lost by accident	9	9	1	12	6	—	4	—
Hatched	17	35	35	60	44	27	27	107
Not hatched	41	17	7	6	6	4	6	16
Total eggs incubated	58§	52	42	66	50	31	33	123
Percent hatched	29	67	83	91	88	87	82	87
Percent not hatched	71	33	17	9	12	13	18	13

* Birds that did not lay as 2-year-olds.
† Birds that laid as 2-year-olds.
‡ The data in this column are from females of unknown age but at least 7 to 14 years old.
§ This figure includes 8 eggs laid by females mated to males their own age; if these eggs are excluded the percentage of hatching would be 34 and not 29.
SOURCE: Richdale, *A Study of a Group of Penguins*, Table 46, p. 5c.

gathered concerning adolescent sterility in the few mammals studied."

It has long been known that the first eggs of pullets tend to be infertile, but Richdale's study constitutes the first thoroughgoing demonstration of the existence of a period of adolescent sterility in birds. It also represents the first proof of the fact that the reproductive development of birds follows, in its general outlines, much the same course as that in mammals.

THE MOUSE

In an experiment reported in 1930 by Mirskaia and Crew on 100 female mice which were mated at first estrus, pregnancy followed in only 24 cases; 11 pregnancies occurring among the 58 colored (19.5 percent) and 13 pregnancies among the 42 albino (30.9 percent) mice. When, however, these same mice were three to six months old they exhibited a fertility ratio, that is, the percentage of pregnancies following matings, of 80 to 90 percent. Yet at first estrus or puberty only 24 percent of matings resulted in pregnancy.

Puberty in the mouse is defined by Mirskaia and Crew as "the stage of individual development characterized by the ability to elaborate functional gametes, and by the physical ability to, and desire to, play the appropriate rôle in mating."

The authors write further: "A review of the facts presented in this paper permits us to attempt a definition of maturity. It is manifest that puberty does not imply maturity of the individual, for when puberty is attained, neither the sex equipment nor the individual are full grown. We suggest that maturity is that stage of individual development which is characterized by the exhibition of the maximum fertility ratio with reference to the conditions of husbandry that are present, and in the case of the female by the ability to produce viable offspring and to rear them. Litter size is not a criterion of maturity, for though it can be shown that in a population litter size increases with the chronological order of the pregnancy up to a maximum, it is as possible to obtain a litter of nine in a pregnancy following a first estrus as to obtain a litter of two in a six-month-old animal. In our stock maturity is reached by the third month of life.

"Puberty marks the completion of those processes of development which equip the individual for fruitful mating; maturity marks the completion of those processes of development which equip the individual for efficient reproduction. The two are distinct in time, for puberty is reached long before maturity is attained. This being so it can be understood that pregnancy is possible at and after the first estrus, but that before maturity is reached such pregnancy will commonly be fruitless or dangerous. It is only when maturity is reached that pregnancy is uneventful. The causes of different rates of development leading to puberty and maturity respectively, and exhibited by different individuals, remain to be determined. It is seen from the tables that some individuals are mature at the time of puberty whilst others do not attain maturity until long afterwards."[2]

As will be seen later, these words can be applied without modification to the development of the human female. The authors' observations are worth discussing here in some detail.

In the first place it is to be noted that the number of pregnancies in the albino mice following upon first mating at the first estrus exceeded by 11.4 percent the number of pregnancies in the colored mice. No explanation is offered by the authors for this phenomenon. Since the nutritional and environmental factors were the same for both albino and colored mice, the suggestion is ventured that a genetic difference in the rate of development of the two breeds is probably responsible for the difference in the expression of fecundity, the albino reaching a higher fecundity potential at first estrus than the colored mice. But however probable, the suggestion requires to be experimentally tested. There is good reason to believe that the rates of development of an organ or system of organs of an animal are genetically influenced, but always, of course, in interaction with the environment. With respect to the human species there exists a large number of complicating factors to be taken into consideration in any discussion of so-called racial fecundity potentials; such factors were long ago cogently discussed by Pearl.[3] Of interest in this connection is a study of homogeneous samples of urban-dwelling black and white women living in the United States, who had been married once only, and were without gynecological disease, never practiced contraception, and who exhibited substantially the same age specific fertility rates.[4]

With respect to mankind we may echo Pearl's cautionary words: "It would seem wise to refrain from attempting quantitative appraisal of racial differences in innate fertility until adequate material is at hand for a critically sound and penetrating analysis of the effects of various extraneous variables, social, economic, and particularly, contraceptive."[5]

Since those words were written it has been abundantly shown that wherever socioeconomic conditions have been equalized age at menarche and fertility rates closely approximate each other in different classes and ethnic groups or so-called "races."[6]

In groups of mice such as those reported upon by Mirskaia and Crew all extraneous factors were as nearly constant as could be; the only variables not so constant being presumably the intrinsic ones of difference in genetic constitution. The available evidence strongly suggests a connection here between the probable genetic difference and the difference in the development of reproductive capacity.[7] It is, however, quite possible that other variables may be involved, but upon the nature of these it is unnecessary to enlarge here. What appears to be likely is that the development of reproductive capacity is genetically regulated, but capable of modification by environmental factors, such as, for example, light and temperature.[8-12] It is interesting that girls who are born blind or lose their sight shortly after birth arrive at menarche earlier than sighted girls.[13,14]

The age at first estrus of these mice was: colored 53 days and albino 34 days. During puberty they were capable of mating, but of such matings only 24 percent of conceptions resulted, more than three-fourths of the matings remaining sterile, and thus strongly suggesting the existence of a normal adolescent sterility interval of some 30 days in more than three-fourths of the animals. But as the authors make clear in this study, at least 24 percent of the animals were fertile upon first mating at estrus. This is an observation of some importance, for it shows that some animals are capable of conceiving at first estrus or shortly thereafter, that the period of adolescent sterility is not necessarily an absolutely sterile one during which conception is impossible. But we shall return to a discussion of this matter shortly.

With reference to Mirskaia and Crew's mice we may, then, define adolescent sterility as *the interval of time from first estrus during which, in a given population of mice, a majority of normally developing mice are incapable of conceiving and remain sterile, while those that do conceive either suffer a high mortality rate or exhibit a high proportion of abortions or miscarriages.*

It is probable that the terms of this definition hold good for all animals, including the human female. Thus, the term "adolescent sterility" may be understood and used in the double sense of functional sterility in relation to the adolescent period of the organism's total life span, or briefly and more clearly as the relative sterility of the adolescent as compared with the fertility of the mature organisms in any given population.

As Mirskaia and Crew stated, "not only is pregnancy rare at time of puberty, but it is also dangerous." They showed that not only were 76 percent of mice sterile at first estrus, but of the 24 percent that did conceive seven, or 29.1 percent, died at or immediately following parturition, while in addition four of the albinos ate their young on the day of birth, such animals as the latter usually dying shortly after devouring their offspring. These figures yielded a death rate for the first estrus mothers of 45.8 percent, a figure extremely and significantly high.

These two facts taken together, namely, the rarity of pregnancy and the high maternal mortality rate among those animals that do conceive at first estrus, suggest a natural developmental unreadiness of the female organism for the processes of reproduction at that time as well as for a varying period beyond it.

In explanation of the uncommonness of pregnancy following first mating Mirskaia and Crew suggested that it is reasonable to assume that "it is because the individual body as a whole is not yet physically fitted for the task though the sex equipment is." That is to say, in the mouse the sex equipment though not full grown at puberty is able to carry out the processes of reproduction, though the body as a whole is not fitted to support the viable termination of such processes. These conclusions were confirmed experimentally by provoking precocious maturity of the sexual system of the immature mouse by the injection of pregnancy urine, without producing any associated maturation of the whole body. Coitus in the case of such animals was not followed by pregnancy until the stage at which the first litter is normally produced.

As further evidence in corroboration of the fact of the reproductive unpreparedness of pubertal mice may be cited the same authors' findings that in primiparae suckling their young the pregnancy rate was only 24.1 percent as compared with 50.0 percent for nursing multiparae.[5] The difference, according to these investigators, was not due to difference in incidence of ovulation associated with postpartum estrus but, they suggested, to a difference between pubertal and adult groups in the level of somatic activity necessary for full reproductive functioning.[15]

We note here that in animals which have thus far been studied, and especially in man, a process of rapid growth of the gonads is initiated during puberty, and that in a relatively short space of time during the period of adolescence they reach almost their adult size.[16,17] Quo has shown that at the onset of puberty the velocity of adult growth in the human female is at its maximum, and he also found that the advent of puberty is preceded by an acceleration followed by a retardation of growth in weight in the two sexes.[18] In the female the musculature, and the form and size of the uterus undergo a rapid series of differentiations, approximating almost to the adult condition.[19-21] Such postpubertal changes in the development of the sexual equipment are chiefly structural. Physiologically the differentiation of the sexual system, most importantly the activities of the hypothalamus and the glands associated with it, the pituitary and the ovary, proceeds at a much slower pace, and it is not until the functions of the latter are properly established that effective reproduction becomes possible.[22-24] At puberty the hypothalamic-axis pituitary cap begins to produce hormonal changes which at first serve chiefly to stimulate the growth of the "sexual equipment," and but slowly serve to initiate those changes in the ovaries which only after some time has elapsed become fully established. It is primarily in the sense of these qualifications that Mirskaia and Crew's statement that "at puberty the individual body as a whole is not yet physically fitted for the task [of reproduction] though the sex equipment is," may be understood.

Additional support for these observations was provided by the experimental work of Engle on the development of puberty and maturity in mice.

Seventy-two immature mice ranging in age from between 17 and 20 days were the subjects of this experiment. Of these animals 63 were given implants of anterior pituitary lobe daily on three successive days, while in the 11 remaining animals the glands were implanted on the first, second, and fourth day of treatment. After the third day of treatment the mice were kept with vigorous mature males. This induced precocious rupturing of the vaginal plate,* at the end

*The vaginal plate is a membrane which in immature rodents closes the vagina. This usually breaks down just before the first estrus. Not to be confused with the vaginal plug.

of the third day or during the fourth day, which in normal mice ruptures at about the thirty-fifth day, and was followed by induced first estrus from one to five days later during which thirty-seven animals mated. From one to five days after mating 30 of these animals were autopsied. Eleven or slightly more than one-third, were found to have ovulated. In some cases where two or three days had elapsed before autopsy no eggs were to be found, the tube being full of detritus as in unmated animals killed after ovulation. In these cases the corpora [lutea] were carefully examined, and those with retained ova were found in numbers only as great as occurs in normal animals. In other cases 15 to 20 morulae or blastocysts were counted in one tube or uterine horn and checked against the number of corpora lutea. In all cases there was considerable discrepancy, showing that many eggs either had not been fertilized or had degenerated following fertilization. Those which had been fertilized were beginning to fragment at the time of examination. On the other hand there were found, in animals killed at the appro-

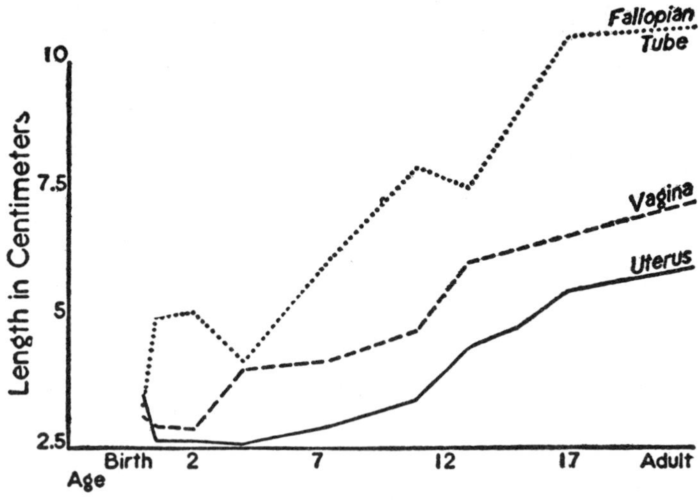

Figure 1. Growth in length of pelvic organs from birth to maturity. (Based on the figures of Scammon after Fluhmann.)

priate stages, many morulae in the tubes, and blastocyst stages within the uterine lumen, yet none were implanted.

"Of the five animals which had been mated more than once, three did not ovulate. Seven animals mated on the first to the fifth day after establishment of the vaginal orifice, but were not sacrificed. If any eggs were fertilized following mating, no animal gave a positive pregnancy sign on the twelfth, thirteenth or fourteenth day, as is constant in pregnant mice."[25]

These observations fairly conclusively proved that while ovulation may be prematurely induced in the immature mouse by anterior pituitary implants, and fertilization of such ova produced, *implantation does not occur,* presumably owing to the absence of the sufficient anterior pituitary hormone which must first act upon the corpus luteum. Even in the animal in which estrus has been precociously induced, the essential remainder of the body is quite unprepared for the function of reproduction.

In the light of these observations Engle made the point that since these animals cannot be considered sexually mature the term "precocious sexual maturity" cannot justly be used to describe the condition induced in them.

Engle's study of the growth of the ovarian follicle in prepubertal mice, in which two distinct phases of growth are noted, from birth to the twelfth day, and from the twelfth day to puberty at the thirty-seventh day, is of interest here as exhibiting the gradual development of the morphological bases for reproduction. On the twelfth day the follicular antrum is seen first. On the thirteenth and fourteenth days this is slightly augmented, while on the fifteenth day there is considerable increase in the size of the antral cavity, almost to the average size of the follicle at first estrus. "From the 15th day to the first estrus the growth of the follicle has quantitatively no important growth change. What changes do occur are those of tissue differentiation."[26]

It was shown in 1930 by MacDowell and fellow workers that the fifteenth day is a crucial period in the life of the mouse.[27]

Engle draws the following conclusions:

"These studies on the ovarian follicle in the 20 days immediately preceding estrus, indicate that these large follicles must be secreting—and liberating—certain amounts of estrin from the 15th day of life, and again emphasize the fact that puberty is a process of slow development, in which the first estrus or first menstruation is merely an incident, expressive of a cumulative action of a delicate hormone balance in the organism. Since the follicles have early attained nearly their full size in less than half the life span before puberty, it is clear that the changes occurring, which cannot be morphologically detected, must be those of slow physiological maturation of the organs which operating satisfactorily together, constitute the living animal."[26]

And again:

"It is evident that while the interaction of the gonads and the anterior lobe affords the physiological mechanism for the first cycle as for the rhythmically succeeding ones, it is evident that this in turn is influenced by somatic factors, and that a general somatic maturity, is essential to a complete reproductive potential."[27]

In this respect Engle's work constituted an independent confirmation of Mirskaia and Crew's work on genital maturity and somatic immaturity.[28] It is worth noting that Engle considered the rupture of the vaginal plate, the first estrus, the first ovulation, and the first mating in the mouse as constituting the several phases of the period of puberty.

The first phase of puberty in the mouse occurs overtly as rupture of the vaginal plate, followed by estrus some 48 hours later—an event which in not more than a third of animals is synchronous with ovulation—and by mating shortly thereafter. It would seem, however, that since first ovulation occurs synchronously with the first estrus in little more than a third of the cases, it would be better to define puberty according to the steplike physiological changes as they appear in the majority of mice, and perhaps to put ovulation after the first mating. The lag between estrus and first copulation is generally only a few days, and in the majority of animals ovulation will normally follow during the next cycles after this. Since it is during these latter cycles that pregnancies occur in a significantly large proportion of cases as compared with matings at first estrus, it would appear reasonable to regard the first ovulation occurring synchronously with first estrus as a precocious development viably unsatisfactory, and to view those cases in which first ovulation occurs at an appreciable interval after first estrus as the more normal condition. In both cases, in any event, it would seem that first ovulation is properly to be regarded as a phase of puberty and not of maturity; maturity being perhaps more correctly regarded as that stage at which the reproductive system is at that level of development which enables it to perform the functions of reproduction in a viable and efficient manner.

MacDowell and coworkers found that the young nulliparous mouse ovary ovulates a smaller number of eggs than the parous and that corpora lutea increase with each successive pregnancy, even up to the tenth.[29,30]

Murray and Bittner demonstrated that first litters of mice are smaller than the second, third, and fourth, and that the litter size drops off after the fourth litter.[31,32] While the pattern of litter sizes is not the same for all strains of mice it appears to be universally the case that in all strains the first litter is smaller than the second.[33] It seems clear that with increasing maturity and parity the maternal mouse becomes more efficient in reproductive capacity. The members of first litters are more susceptible to disease and weigh less than those of later litters. It has also been noted that the mother's milk increases with successive litters.[33]

Runner and Gates found that when conception is experimentally produced in prepubertal mice uterine implantation occurred in less than 10 percent of females. These investigators took 37 female mice approximately 35 to 40 days old and gave them a priming injection of 1 IU (international unit) of pregnant mare's serum 60 hours before desired ovulation. Some 13 hours before desired ovulation an injection of 1 IU of human chorionic hormone was given.

Each female was caged with an adult male. The next morning 34 females showed evidence of having mated—each exhibited the vaginal plug.* Fifteen of the animals examined eight days after ovulation showed no indications of uterine implantation. In other experiments the investigators found evidences of uterine implantation in less than 10 percent of the animals.[34]

In the course of a discussion of adolescent sterility in the human female Mills and Ogle cited evidence that they considered to support the existence of an adolescent sterility period in the human species. They stated that such evidence would not be considered conclusive "without corroboration on experimental animals under controlled conditions. We have no records of timing of intercourse in . . . human cases . . . hence much of the discussion must remain problematical. It so happens, however, that animals give definite evidence of this same adolescent period of sterility. . . ."[35]

The suggestion that the findings on mice may be used to "corroborate" the existence of an adolescent sterility period in humans and render the evidence for such a period conclusive, is open to some question. Mice and humans are two very different systems of variables, and sequences of facts observed in connection with the one cannot safely be generalized to hold good for the other. The most that can be said here is that since mice and humans are mammals, and since in a general way their bodies grow and function in a similar manner, it is possible that, allowing for all the differences in rates of growth and development, a similar infecundity period characterizes the early adolescent human individual. As Hartman puts it, "what is phylogenetically significant is apt to be physiologically fundamental."[36]

Such a possibility would be tantamount to a guess were it not for the fact that in all the animals thus far investigated the facts all point in the same direction. It is, however, one

*Vaginal plug formed from seminal fluid coagulated by a prostatic enzyme. The plug seals the vaginal orifice.

thing to speak of "corroboration" and the rendering "conclusive" of evidence projected from one distinctive group to the uncertain evidence of a very different one. Yet when such cautionary comments have been made the fact remains that such general conclusions as have been drawn by Mirskaia and Crew with respect to the development of puberty and maturity in the mouse may, with the time element extended, be applied almost word for word to humans.

The evidence Mills and Ogle's study corroborated and rendered conclusive was that of Mirskaia and Crew, although they did not refer to the work of the latter investigators.

Mills and Ogle were interested in the relation between climate and the onset of menarche, and in the effect of climate upon the fertility of man; in order to learn something of the nature of this relation they resorted to white mice as their experimental animals. The account of their experiment and their results are best set out in their own words:

"Mice grown from a single stock were divided into four groups. One group was kept in the control breeding room at 70 to 80° F., another was placed in a special room kept at 64 to 68° F., and a third placed in a hot room kept at 88 to 92° F. and about 70 percent relative humidity. A fourth group was shifted daily between the hot and cold rooms, spending 12 to 16 hours of each day in the moist heat. In every case pregnant mothers were allowed to deliver their young in the control room and stay there with the litter until the young were 21 days old. This was necessitated by the fact that the mothers usually kill the young if left in either the heat or cold.

"From Table 2 we see the first estrus, corresponding to the human menarche, beginning at 33 days of age in the control room group, at 36 days for the young of control room mothers placed in the cold room at 20 days of age, at 33 days for the cold room young of cold room mothers, and at 35 days for the young of control room mothers shifted daily from hot to cold. Young placed in the moist heat, however, show a definite and striking delay in the age of first

Table 2
First Estrus and First Delivery Data on Whice Mice

Description of Group	No. of Cases	Average Age First Estrus (days)	Age First Delivery (days)	Lag in Fertility First Estrus to First Conception (days)
Breeding room controls	20	33	84	31
Cold room immigrants	11	36	88	32
Cold room descendants	25	33	79	26
Hot room immigrants	31	37	94	37
Hot room descendants	9	46	—	—
Change group immigrants	26	35	138	83

ADAPTED from Mills and Ogle.[35]

estrus. Young of control room mothers, placed in the heat at 20 days of age, began the oestrus at 37 days of age, but the young of the hot room mothers were delayed to *46 days,* confirming our earlier statements on the later menarchial age for girls of native tropical stock.''[37]

From these observations the following facts emerge:

(1) only 5 percent of each of the groups of mice conceived at each of the first three estrus periods, and it was not until the sixth and seventh estrus periods that full fecundity was established; (2) the earlier the appearance of the first estrus, the longer was the lag in fertility; (3) the later the appearance of first estrus, the shorter was the absolute lag in fertility; (4) the average lag in fertility for three normal control groups of mice was 31.8 days; (5) stimulating cold produced a quickening in sexual development; (6) moist heat produced a depression and a slower development of the sexual functions, and an increase in the fertility lag; (7) great temperature variability produced a marked suppression of fertility.

We may conclude, then, from the work of Mirskaia and Crew and of Mills and Ogle, on different stocks of mice under very different conditions, that the majority of mice are characterized by an adolescent sterility period of some 31.8 ± 1.9 days.

THE GOLDEN HAMSTER

Diamond and Yanagimachi found that the golden hamster ovulates at 27 days after birth and that first matings occur at 43 to 46 days. Earliest successful pregnancies occur at 50 to 70 days. The adolescent sterility interval in the golden hamster therefore varies from seven to 26 days, with a mean of 16 days. This is a very striking fact in a species which develops sexually so early.[38]

THE SHEEP

Dr. H. Goot of the University of New Zealand School of Agriculture, Palmerston, North, has kindly sent me data on sheep that indicate the existence of an adolescent sterility interval in the sheep. The data contained in Tables 3 and 4 express the lower fecundity rates in the younger ewe as compared with the mature ewe, the decline in fertility setting in with the reaching of the age of five and one-half years and over.* Matthews Duncan as long ago as 1866 quoted two breeders to the effect that "the mortality of the young of

Table 3
Infertility Among Dry Ewes

Age at Mating (in years)	%* Once-twice Dry	%* Once Dry	%* Twice Dry†
1.5	16.6	16.6	0.0
2.5	11.2	6.3	4.9
3.5	5.8	3.9	1.9
4.5	5.9	3.3	2.6
5.5 years and over	6.0	3.5	2.5
Total and average	10.2	8.0	2.2

SOURCE: Goot 1949. Personal communication.
NOTE: Dry ewes are barren and/or ewes that aborted.
* Based on number of ewes put to rams.
† All twice-dry ewes are culled.

*H. Goot, 3 October 1949, personal communication.

Table 4
Mortality Among Lambs

Age of Ewes at Mating (in years)	Lambs Born	% Died Before Docking†	Lambs Died before Docking			
			No. Singles Born	No. Singles Born Dead	No. Twins Born	No. Twins Born Dead
1.5	2,270	16.3	373	52 (13.9)‡	254	55 (21.7)
2.5	1,967	10.4	230	18 (7.8)	286	32 (11.2)
3.5	1,862	10.3	162	19 (11.7)	312	33 (10.6)
4.5	1,380	10.8	106	14 (13.2)	206	29 (14.1)
5.5 and over	1,704	10.5	92	13 (14.1)	254	46 (18.1)
Total and average	9.183	11.9	963	116 (12.1)	1312	195 (14.9)

SOURCE: Goot 1949. Personal communication.
* One flock only.
† Based on lambs born.
‡ Numbers in parentheses indicate percentages.

these animals [lambs and calves], when the mothers are immature, is much greater than when they are well grown. . . . One of them says, 'Taking the first lamb from ewes at one year old has in almost every case failed to be remunerative, owing to the frequent deaths of the lambs. The same may be said of young heifers though the mortality of the offspring may not be so marked as in that of sheep.' "[39]

THE COW

In a study of the prenatal development of the cow, Winters, Green, and Comstock found that cows bred at first or second estrus yielded an exceptionally high proportion of abnormal ova. "The recovery," they write, "of this relatively large proportion of abnormal eggs from these heifers bred during their first or second heat period helps to explain observed 'sterility' encountered in young females."[40]

Professor L. M. Winters, Professor in Charge of Animal Genetics at the University of Minnesota Agricultural Experiment Station, has informed the writer that in perfectly healthy heifers bred in the first heat period the percentage of eggs and early embryos that were recovered in varying stages of decomposition was considerably higher than would be expected in a group of healthy mature females. Professor Winters states he has always attributed these conditions to the immaturity of the heifers.*

Attention may here be drawn to the fact that at least as early as the seventh century B.C. it was already known that:

The premature offspring is always thin:
A cow's first heifer is inferior,
Her second offspring is twice as large.

as these lines from the Babylonian poem known as "The Babylonian Ecclesiastes" testify.[41]

It seems clear that in the penguin, mouse, and appar-

*L. M. Winters, 17 July 1939, personal communication.

ently in the sheep and cow, experiment and observation show that the adolescent, immature animal is characterized by a variable period of relative sterility. The majority of such animals are incapable of conception at or shortly after first estrus, most are incapable of conception until a goodly number of estrus cycles have been experienced, while the small proportion of those animals that are capable of conception at these times are characterized by high fetal, infant, and maternal death rates. In addition, in the cases of the mouse and cow, there is definite evidence of a large proportion of abnormal fertilized and unfertilized eggs, and decomposing early embryos.

It is probable that these findings may be generalized for most, if not for all, animals.

REFERENCES

1. Richdale, L. E. A study of a group of penguins of known age. Dunedin, New Zealand: *Otago Daily Times*, 1949.

2. Mirskaia L., and Crew, F. A. E. Maturity in the female mouse. *Proc R Soc Edinburgh*. 50: 179–86, 1930.

3. Pearl, R. Biological factors in fertility. *Ann Am Acad Political Social Sciences*. November, 1936.

4. Pearl, R. *The Natural History of Population*. New York: Oxford University Press, 1939.

5. Pearl, R. Fertility and contraception in urban whites and negroes. *Science*. 83: 503–6, 1936.

6. Zacharias, L., and Wurtman, L. J. Age at Menarche. Genetic and environmental influences *N Engl J Med*. 280: 868–75, 1969.

7. Crew, F. A. E., Puberty and Maturity. *Proc 2nd International Cong Sex Res*. London: Oliver & Boyds, 1931.

8. Bissonnette, T. H. Sexual photoperiodism. *Q Rev Biol*. 11: 371–86, 1936.

9. Mills, C. A. Geographic and time variations in body growth and age at menarche. *Hum Biol*. 9: 43–56, 1937.

10. Mills, C. A. Climatic effects on growth and development, with particular reference to the effects of tropical residence. *Am Anthropol*. 44: 1–13, 1942.

11. Montagu, Ashley. Climate and reproduction. *Science*. 89: 290–2, 1939.

12. Montagu, Ashley. *An Introduction to Physical Anthropology*. 3d Ed. Springfield, IL.: Charles C Thomas, 1962.

13. Zacharias, L., and Wurtman, R. Blindness: its relation to age at menarche. *Science*. 144: 1154–55, 1964.

14. Zacharias, L., and Wurtman, R. Blindness and menarche. *Obstet. Gynecol.* 33: 603–9, 1969.

15. Mirskaia, L., and Crew, F. A. E. On the pregnancy rate in the lactating mouse and the effect of suckling on the duration of pregnancy. *Proc R Soc Edinburgh*. 51: 1–7, 1931.

16. Ingalls, N. Sex features in adolescence. *Proc Conf Adolescence*. 1930.

17. Greulich, W. W. et al. Somatic and endocrine studies of pubertal and adolescent boys. *Monographs Soc Res Child Devel.* 8: 3, 1942.

18. Quo, S. K. Mathematical analysis of the growth of man with special reference to Formosans. *Hum Biol.* 25: 33–358, 1953.

19. Bolt, R. A. Growth of the muscles during the second decade. *Proc Conf Adolescence*. 1930.

20. Dorfman, R. I., Greulich, W. W., and Solomen, C. I. The excretion of androgenic and estrogenic substances in the urine of children. *Endocrinology*. 21: 741–43, 1937.

21. Oesting, R. B., and Webster, B. The sex hormone excretion of children. *Endocrinology*. 22: 308–14, 1938.

22. Gundobin, N. P. *Die Besonderheiten des Kindesalters*. Berlin, 1912.

23. Scammon, R. E. The growth of the human reproductive system. *Proc 2nd International Cong Sex Res*. London, 1931.

24. Fluhmann, C. F. *The Management of Menstrual Disorders*. Philadelphia: W. B. Saunders, Co., 1956.

25. Engle, E. T. The pituitary-gonadal relationship and the problem of precocious sexual maturity. *Endocrinology*. 15: 205–420.

26. Engle, E. T. Prepubertal growth of the ovarian follicle in the albino mouse. *Anat Rec*. 48: 341–50, 1931.

27. MacDowell, E. C., Gates, W. H., and MacDowell, C. G. The influence of the quantity of nutrition upon the growth of the suckling mouse. *J Gen Physiol.* 13: 529, 1930.

28. Mirskaia, L. and Crew, F. A. E., On genital maturity in an immature soma. *J Exp Biol.* 7: 248–52, 1930.

29. MacDowell, E. C., and Lord, E. M. The number of corpora lutea in mice. *Anat Rec*. 31: 131–41, 1925, 1929.

30. MacDowell, E. C., Allen, E., and MacDowell, C. G. The relation of parity age, and body weight to the number of corpora lutea in mice. *Anat. Rec.* 41: 267–72, 1929.

31. Murray, W. S. The breeding behavior of the dilute brown stock of mice. (Little dba). *Am J Cancer.* 20: 573–93, 1934.

32. Bittner, J. J. Differences observed in an inbred albino strain of mice following a change in diet. I Litter size. *Nutr Bull Jackson Mem Lab.* 1: 3–9, 1936.

33. Russell, E. S. Search for new cases of parental and seasonal influences upon variations within inbred strains. *Ann NY Acad Sci.* 59: 597–605, 1954.

34. Runner, M. N., and Gates, A. Conception in prepuberal mice following artificially induced ovulation and mating. *Nature.* 174: 222–3, 1954.

35. Mills, C. A., and Ogle, C. Physiologic sterility of adolescence. *Hum Biol.* 8: 607–15, 1936.

36. Hartman, C. G. The corpus luteum and the menstrual cycle together with the correlation between menstruation and implantation. *Am J Obs Gyn.* 19: 511–22, 1930.

37. Mills, C. A., and Ogle, C. Physiologic sterility of adolescence. *Hum Biol.* 8: 612–3, 1936.

38. Diamond, M., and Yanagimachi, R. Reproductive development in the female golden hamster in relation to spontaneous estrus. *Biol Reprod.* 2: 223–9, 1970.

39. Duncan, J. Matthews. On the age of nubility. *Edinburgh Med J.* 12: 207–14, 1866–67.

40. Winters, M., Green, W. W., and Comstock, R. E. Prenatal Development of the Bovine. *Tech Bull* 151. St. Paul, Minn.: University of Minnesota Agricultural Experimental Station, 1942.

41. Pritchard, J. B., editor. *Ancient Near Eastern Texts.* Princeton, N. J.: Princeton University Press, 1950.

3

THE RHESUS MONKEY AND THE JAVA MACAQUE

The rhesus monkey, *Macaca mulatta* (formerly known as *Macacus rhesus*), is yet another animal which has been the subject of investigations that can be safely utilized in discussing the question of adolescent sterility.[1] These investigations were carried out originally by C. G. Hartman on the Carnegie colony of animals at Baltimore, and first briefly reported in 1931.

In his 1931 communication Hartman adduced evidence to show that the generally accepted belief that puberty or maturity or reproductive capacity are synonymous or synchronous phases of development is erroneous. From his examination of the evidence Hartman was led to the conclusion that the interval of time that normally elapses between puberty and maturity in man, monkey, and rat "is about what one would expect with the application of Donaldson's law of equivalent ages: three years, one year and one month respectively, for man, monkey, and rat."[2]

Reporting his findings on fifteen rhesus monkeys, whose menstrual cycles had, from inception, been carefully observed, Hartman wrote: "These fifteen females averaged 3,350 grams in weight at first menses. Most of them were mated soon after puberty, but not a single female conceived before attaining a weight of 4,370 grams. The average weight at first conception of nine of these females was 5,000 grams; hence it is apparent that many menstrual cycles passed between the first menses and the first conception, despite frequent matings in the interval. The interval may be estimated at about a year of time."[2]

In these young animals Hartman found that "the menstrual cycles of the young females are extremely irregular in length and duration, and there is a very low incidence of ovulatory as compared with nonovulatory cycles."[2]

In his monograph on the reproduction of the rhesus monkey published in 1932 Hartman made it clear that the failure to ovulate was "the probable cause of the relative sterility of adolescent females."[3]

In his earlier communication of 1931 Hartman went on to suggest that similar conditions characterize the human female, and that although the mores of a given people may force effective marriage upon the girl at her first menstruation, "there is much indication that, by and large, nature herself prevents motherhood supervening during an important series of preparatory years."[2]

The greater part of the communication is, in fact, devoted to the establishment of a case for the existence of a period of relative sterility in the adolescent human female, the concluding statement of which is to the effect that "It seems highly reasonable that Malinowsky's [*sic*] predicament is explained by the facts presented in this paper, namely, that the first menstruation (*puberty*) marks merely an early manifestation of a train of events (*adolescence*) which only after three or four years on the average lead to ovulation and conception, the proof of *maturity*."[2]

Hartman points out that the terms adolescence, puberty, and maturity are often used loosely and interchangeably, and for the purposes of his own study defines these terms as follows: "*Puberty* is used synonymously with the menarche, the time of the first menstruation; *maturity*, the age of the first ovulation and the first uterine preparation for pregnancy; *adolescence,* the two periods combined, that is, the entire period when reproductive functions are in the process of attaining the climax, reproductive capacity, namely, ovulation, fertilization, implantation. For there is reason to believe that adolescence is not a sudden but a gradual change in the life of the individual mammal and that in primates the first ovulation is separated by months and perhaps years from the first menstruation.

"Some observations on the monkeys of the Carnegie Colony have led to these conclusions. In the first place no adolescent family has conceived during the first two years after puberty. Furthermore, there is a very high incidence of nonovulatory cycles in young animals. These observations point to the failure to ovulate as the probable cause of the relative sterility of adolescent females."[3]

It will be recalled that in his earlier communication Hartman gave the interval between menstruation and first conception in his rhesus monkeys as "about a year of time." In his 1932 monograph he states that "no adolescent female has conceived during the first two years after puberty." A careful study of Hartman's protocols indicates that the first estimate of about one year is the correct one, and in a personal communication Dr. Hartman has been kind enough to confirm this.

In summarizing the evidence Hartman concludes: "Adolescence is a period of endocrine instability, as evidenced by irregular menstruation, preponderance of menstruation without ovulation, follicular atresia, excessive swelling of the sex skin, comparative sterility. The sterility of adolescence is due to failure to ovulate. This, of course, entails a failure of the uterus to undergo the pregravid changes in preparation for the reception of an egg. It is highly probable that the uterus 'learns' to undergo these changes in a step-like fashion, the organ growing a little larger with each succeeding cycle in which the ovary likewise produces a larger and larger follicle until this no longer degenerates but goes 'over the top' and ovulation occurs."[3]

Smith and Rubenstein, writing in 1940, report their observations of the reproductive cycles in six adolescent female rhesus monkeys for a period of nine months, their purpose being to describe the period between menarche and conception, the period of adolescent sterility in terms of ovarian function.

During the premenarcheal period sexual skin activity was observed in only three of the animals, in another it commenced in the intermenstrual period following menarche. The investigators were able to determine that "the first menstrual cycles were unfailingly anovulatory. Small follicles were continually and persistently palpable on the ovaries of each of the monkeys during the intermenstrual period. The vaginal smears showed very high cornification. Nevertheless, ovulation failed to occur during this initial cycle, and probably for the same reason that in the smears the cornification remained constantly high. There was no aggregation and folding of cells; there was very little desquamation; all of which indicated a lack of luteinization."[4]

In two macaques ovulation had not occurred five months and nine months respectively following menarche. The authors suggest that these intervals represent not so much periods of adolescent sterility as times of gonadal inactivity and regression due to poor health. In a third macaque the first four postmenarcheal cycles were anovulatory. When a follicle was palpated and found to have ruptured the animal was immediately placed with a male and successfully impregnated. The pregnancy ended in a spontaneous abortion after 82 days.

Three macaques conceived during the second cycle following menarche. In one of these monkeys there was a spontaneous abortion after 20 days, in another after 37 days, while the third carried her thriving fetus successfully to term. It is significant that both in weight (3,900 grams) and skeletal development at conception this animal was the most advanced.

Thus, as the authors point out, "four ovulations and four conceptions following brief periods of sterility have been observed among the six adolescent macaques. Three out of the four conceptions resulted in spontaneous abortions. Therefore, it would seem that the animals were still too immature to reproduce, although their immaturity was expressed not in failure of ovulation but perhaps in inefficiency of the uterus in maintaining gravidity."[4]

The authors conclude, "The data indicate that puberty in the female macaque is followed by gradual physiological development of the genital and endocrine systems until the two are capable of working synchronously and successfully at the function of reproducing. The length of time required to complete this development differs for individual animals but has not been less than a month and seldom more than a year. The interval is a time of anovulatory and irregular menstrual cycles. . . . The high incidence of anovulatory cycles is probably due to the failure of the pituitary to produce sufficient luteinizing hormone. . . . The cases of early pregnancy which ended in spontaneous abortion indicate that the adolescent female macaque may be unprepared for reproducing even after the animal is capable of ovulation and conception. The animal which carried a normal fetus to term, having conceived one month after menarche, proves that sterility in the adolescent is a relative phenomenon dependent upon the general health, and the tempo of the growth processes throughout the body for the extent of its duration."[4]

Confirming Hartman's observation, Smith and Rubenstein found that menarche in the female macaque occurs at an average weight of 3,400 grams, and they conclude that "the sterile period of the adolescent monkey may be but one month, or it may continue for a year or more depending on the health and growth rate of the animal."[4]

Hotchkiss, Atkinson and Knobil have shown that ovulation during the menstrual cycle is initiated in adult rhesus monkeys by a surge of luteinizing hormone (LH) and follicle stimulating hormone (FSH), preceded by a rise in circulating estrogen levels. This increase is the primary stimulus for the preovulatory discharge of gonadotropic hormone. Positive feedback action of estrogen on LH and FSH secretion in young female rhesus monkeys that are chronologically near menarche does not occur. Dierschke and his coworkers have found that although the positive feedback system is not operative in most monkeys during the first few months after menarche, it can usually be demonstrated shortly before spontaneous ovulatory cycles are detected. The latter event, these investigators have found, does not generally occur until about a year after menarche.[5,6] Following a variable number of irregular anovulatory cycles, the circulating levels of estrogen rise above the threshold for triggering surges of LH and FSH, by which time the system has become competent to respond to the positive feedback action of estrogen; ovulation, and corpus luteum formation ensue and regular, ovulatory cycles predominate thereafter.[7]

Dr. John Resko of the Oregon Regional Primate Research Center, at Beaverton, Oregon, has been kind enough to supply me with some unpublished observations on the rhesus monkey (*Macaca mulatta*). He has taken serial blood samples from a small number of females and analyzed the plasma for concentrations of progesterone that would indicate whether or not ovulation had occurred. In these laboratory females menarche first occurred around 30 months of age, and none of them ovulated during their first three menstrual cycles. Approximately one-third then ovulated during the next two or three cycles, another third ovulated during the following two or three cycles, and almost all of the rest of the females ovulated by the twelfth cycle. It is not known whether these were fertile ova.* What is of interest here is the physiological demonstration of the fact that the early menstrual cycles in the rhesus monkey are anovulatory.

Thus early in our examination of the evidence for the primates the indications are, and our later analysis of the

*John Resko, 12 October 1977, personal communication.

physiological facts for the remainder of the primates including man will show, that adolescence is a lengthy period of increase in functional complexity marked by the slow development of two primary phases. These are the phase of *puberty,* which begins with menarche or the first menstruation, and the phase of *nubility,* the period at which the female becomes capable of procreation without appreciable danger to herself or to her offspring. With the termination of the period of nubility, the female may be said to reach *maturity,* the period during which the female is at the optimum of her procreative capacity.

Spiegel, in a series of studies on the Java macaque (*Macacus cynomolgus*), has demonstrated that this species exhibits a sexual development similar to that of the rhesus monkey.[8-10]

In the eleven female monkeys studied the average age of appearance of the sexual skin was two years and six months. Menarche appeared at an average age of two years and 10 months, and first conception, after being accessible to mature males, occurred in nine animals at an average age of four years and four months. In two animals observed up to the ages of six years and nine months in the one case and seven years and eight months in the other, conception had not yet occurred. Omitting these two animals on the suspicion of possible abnormality, the average age at conception of four years and four months yields a sterility interval for the remaining seven animals of one year and six months.

In the male Java macaque Spiegel found that the average age of "puberty" as determined by the presence of the first spermatozoa in the ejaculate in six animals was three years and nine months, and the average age at first fruitful insemination was four years and two months, yielding an adolescent sterility interval of about five months.

REFERENCES

1. Corner, G. W. The reproductive cycle of the Rhesus monkey. Edited by G. A. Baitsell. In *Science Progress.* New Haven, Conn.: Yale University Press, 1951, 27–58.

2. Hartman, C. G. On the relative sterility of the adolescent organism. *Science*. 74: 226–7, 1931.

3. Hartman, C. G. Studies in the reproduction of the monkey *Macacus (Pithecus) rhesus*, with special reference to menstruation and pregnancy. *Carnegie Institution of Washington, Contributions to Embryology*, No. 134, 1932, 23.

4. Smith, R. M., and Rubenstein, B. Adolescence of macaques. *Endocrinology*. 26: 667–9, 1940.

5. Dierscke, D. J., et al. Hypothalamic pituitary regulation of puberty: Feedback control of gonadotropin secretion in the Rhesus monkey. Edited by W. M. Grumbach, G. D. Grave, and F. E. Mayer. In *The Control of Puberty*. New York: John Wiley & Sons., 1974, 104–14.

6. Grumbach, W. M., Roth, J. C., Kaplan, S. L., and Kelch, R. P. Hypothalamic-pituitary regulation of puberty in man: Evidence and concepts derived from clinical research. Edited by W. M. Grumbach, G. D. Grave and F. E. Mayer. In *The Control of Puberty*. New York: John Wiley & Sons, 1974, 155.

7. Dierschke, D. J., Weiss, G., and Knobil, E. Sexual maturation in the female rhesus monkey and the development of estrogen-induced gonadotropic hormone release. *Endocrinology*. 94: 198–206, 1974.

8. Spiegel, A. Beobachtungen über den Sexual Zyklus, die Gravidät und die Geburt bei Javamakaken. (*Macaca irus mordax*) Thomas und Wroughton [cynomolgus L.]. *Archiv für Gynäkologie*. 142: 561–591, 1930.

9. Spiegel, A. Untersuchungen über die Fortpflanzung bei Javamakaken. *Zentralblatt für Gynäkologie*. 55: 227–70, 1954.

10. Spiegel, A. Beobachtungen and Untersuchungen an Javamakaken. *Zoologisches Garten,* N. F. 20: 227–70, 1954.

4
THE CHIMPANZEE

The only other animal for which a fair amount of reliable information is available with reference to the problem of adolescent sterility is, fortunately, a close collateral relative of man, namely, the chimpanzee.

Tinklepaugh in 1933 reported a chimpanzee, in whom the first menstrual bleeding was observed when she was at an estimated age of nine years. The first swelling of the sex skin had appeared five months previously and there were four cycles of sex skin swelling prior to the establishment of menarche. She was continuously caged with a mature male, but conception did not occur until four months after first menstruation. It is not without significance that this mother died 15 days after giving birth.[1]

She may have died of puerperal septicemia or from the inability to withstand the strain of reproduction upon her system. Recalling the fate of early-conceiving mice it would look as though she had possibly conceived too early in her development for her own good, and in this respect her history would correspond to the early-conceiving mice of Mirskaia and Crew, almost half of whom either died or ate their young. This is rendered all the more probable when we consider the following two well-authenticated cases.

Yerkes in 1935 reported on a second generation captive-born chimpanzee, Cuba, as follows: "She first exhibited characteristic genital swelling [edema of the sexual skin] in April 1933. Menstrual bleeding occurred first on July 10, 1933, when she was seven years, four months, old. She was caged with a mature male from May 1933, and she became pregnant August 9 (± 5 days), 1934, at the age of eight years, five months."[2]

The sterility interval in this chimpanzee was, then, one year and one month. It is further to be noted that in this case parturition, which occurred on 11 April 1935, was uneventful and that the mother and infant both did well, the infant weighing close to four pounds at birth.

"Mother and infant were observed continuously from 3:20 to 4:10 P. M., and both verbal and pictorial records were made of their behavior. As a primiparous mother Cuba exhibited behavioral inadequacies. Although from the first she carried Peter about with her, holding him awkwardly, usually grasped in one hand, she did not, according to species practice, place him upon her abdomen or breast and permit him to cling to her. Instead she treated him much as she might any strange object which interested, puzzled and annoyed her. Toward the end of the period of observation she forcibly broke his hold upon her whenever he succeeded in grasping her hair or skin with hand or foot. Often in so doing she was rough and impatient and vocalized complainingly.

"Peter was left with his mother for about eighteen hours (overnight) under intermittent observation. As Cuba did not accept him, to be nursed and generally cared for, it was necessary to take him from her, in order that he might not become the victim of her inexperience, curiosity, neglect or abuse."[2]

One may well raise the question whether it is possible that these inadequacies are in some significant way, apart from the factor of experience, associated with the degree of reproductive maturation of the animal. Or is the factor of experience perhaps the important thing here?*

With respect to Cuba, Dr. J. H. Elder has informed the writer that this animal has exhibited somewhat similar behavior with her second infant; although she kept it by her she did not, however, care for it in a satisfactory manner, so that after some seven weeks it had to be taken from her in view of its serious undernourishment.

In the Johns Hopkins chimpanzee, Evo, discussed in 1935 by Schultz and Snyder, "The sex skin† became oedematous for the first time when she possessed all her permanent teeth except the third molars, and at the carefully estimated age of 9 years and 1 month. . . . This swelling persisted for nearly 8 months, i.e., until a few days before her first menstruation.

"The first menstruation of Evo occurred at the estimated age of 9 years and 9 months, and 9 months before the eruption of the lower third molars."[3]

Evo was continuously together with a mature male, from an early age, although intercourse did not take place until after her seventh menstrual cycle or approximately eight months after her first menstruation, when after "coitus between the two was continued fairly regularly 3 and more times daily for approximately 6 months or for several weeks after she became pregnant.[3]

". . . in spite of regular menstruation on the part of the female and abundant coitus on the part of both sexes, no conception took place for a considerable time, namely, for at least 5 months after the first possible exposure and until after the 11th menstrual period, which occurred exactly 1 year after the first menstruation. Evo's estimated age at the

---

*For a further discussion of this matter see pp. 191–194 of the present volume.

†The "sex skin" or "sexual skin" is a massive pinkish swelling of the skin of the perineum and external genitalia under periodic hormonal control. Among the anthropoid apes it occurs in the chimpanzee alone.

time of conception was 10 years and 9 months, or 3 months after her lower third molars had appeared, and about 8 months before her permanent dentition became completed by the eruption of the upper third molars.''[3]

It appears, then, that in Evo the sterility interval lasted for one year. On the 188th day (after the onset of the last menstruation) Evo's health changed for the worse, and on the 196th day of pregnancy, not surprisingly, she spontaneously aborted a stillborn fetus. She subsequently made a good recovery.

Here once more is the case of a first pregnancy in an adolescent chimpanzee that terminated unsuccessfully, and again the evidence suggests that pregnancy in the immature animal is a hazardous process, with high risks for both offspring and mother. We may recall here McLaren's report of the only two native girls who in all his eight years at Cape York were the only ones to give birth before they were married, their offspring being stillborn and undeveloped (see Chapter 1).

Concerning the two other chimpanzees in the Johns Hopkins colony the authors write:

"In the second female Hopkins chimpanzee, Mae, the first sex swelling appeared at a dental age closely corresponding to that in Evo [i.e., about 9 years and 1 month], but, in contrast to that in Evo, this swelling lasted for only about 3 weeks, and has not reappeared at the time of writing, about 5 weeks later. In the third female Hopkins chimpanzee, Pandora, the sex swelling had made its first appearance before the permanent canines had erupted and persisted very pronouncedly for approximately 14 months. Neither Mae nor Pandora has as yet menstruated.''[3]

The authors do not tell us whether Mae was ever exposed to the mature male, but they state that at "about the time of Evo's first menstruation the younger Hopkins female, Pandora, was added to the colony. Dayton had regular and frequent sexual intercourse with Pandora from the start, even though she never menstruated but showed merely pronounced sex swelling throughout the period from her arrival to the present.''[3]

Dr. J. H. Elder kindly supplied the writer with data concerning the reproductive history of the chimpanzee, Soda, which has been under observation for some years at the Yale Laboratories of Primate Biology, Florida. Beginning about the first of June 1935, at an estimated age of seven years and 11 months, Soda showed very slight enlargement of the external genitalia, which persisted with almost imperceptible variations until 19 August, when it disappeared entirely. The swelling began again on 22 September, and this time resembled more closely the typical swelling, although at this time it was still much smaller than those exhibited later. First menstruation occurred 18 November 1935, at an estimated age of eight years and four and one-half months. Conception occurred on the nineteenth day of the seventh cycle, on or about 11 September 1936. Since, however, there were only four opportunities for conception in the cycles previous to the seventh, it is not possible to say with certainty that this animal did not ovulate or was incapable of conception before the seventh cycle. But since on these four previous occasions she was mated with a mature male at the optimum midinterval period for conception, there is a good probability that she had not previously ovulated or, at least, that she was not previously capable of conception. The menarche-conception interval in this case may, then, tentatively be said to have been 10 months.

From histories such as those given above it seems probable that the chimpanzee is, in most cases, normally incapable of conceiving until some appreciable time after the establishment of menarche.

Two additional records of uncertain value may be noted here.

Fox, in an extremely interesting paper written in 1929, on births observed in two anthropoid apes, reported on a chimpanzee, Marianne, who had come to the Philadelphia Zoo in 1924, when she was estimated to be about three years old.[4] She first began to menstruate in 1926 at the age of five years, but it was not until she was seven and a half years old that conception took place. During the interval between menarche and conception she was caged with a male chimpanzee of her own age. Schultz and Snyder state that "a male was not continuously present during this time," but this is not so according to Fox. Intercourse, according to him, took place frequently between the animals during this period. In a personal communication (15 March 1938) Dr. Fox writes: "The companionship in the same cage of the male and female chimpanzee was uninterrupted from menarche to conception. That is what can be read from my article of 1929. . . . They came together and grew up together. There is no reason to think that any period of time passed when they were separated so that from the institution of menstruation of the female chimpanzee until conception the two must have experienced the same physiological development of sexual maturation together and each under the influence of the other."

Since the somewhat younger male may have been in process of maturing during a part of the time during which the female had already attained to reproductive capacity, it is difficult to draw any definite conclusion from this case.

Similarly for the chimpanzee, Boo-Boo, of the London Zoological Gardens, reported by Wyatt and Vevers, the evidence is not quite clear.[5] Boo-Boo first menstruated at an estimated age of eight years in 1931. She was not, however, put together with a mature male until three years later in June, 1934, whereupon she promptly conceived, being at an estimated age of 11 years at the time of conception.

These various observations on the chimpanzee yield four cases in which the conditions necessary to enable one to make a fairly satsifactory judgment concerning the existence or not of a period of adolescent sterility in the chimpanzee have been more or less adequately fulfilled. These cases tabulated according to each investigator's presentation of the data are summarized in Table 5.

Upon the basis of the evidence provided by these four cases it is possible to say that the degree of probability that the chimpanzee is normally characterized by a period of

Table 5
Developmental Sexual History with Special Reference to the Adolescent Sterility Interval in the Chimpanzee

Observer	Animal	First Sexual Skin (date, age)	Sexual Skin Swellings Inception to Menarche	Menarche (date, age)	First Conception (date, age)	Delivery (date, age)	Adolescent Sterility Interval
Tinklepaugh 1933	Dwina	May, 1929 8 yr 8 mo	4	Sept. 8, 1929 9 yr	Jan. 8, 1930 9 yr 4 mo	Sept. 11, 1930 10 yr	4 mo
Yerkes 1935	Cuba	Apr., 1933 7 yr 1 mo	3?	July 10, 1933 7 yr 4 mo	Aug. 9, 1934 8 yr 5 mo	Apr. 11, 1935 9 yr 1 mo	1 yr 1 mo
Schultz and Snyder 1935	Evo	9 yr 1 mo	Continuous until two weeks before menarche	9 yr 9 mo	10 yr 9 mo	—	1 yr 0 mo
Elder 1938	Soda	June 1, 1935 7 yr 11 mo	2	Nov. 18, 1935 8 yr 4½ mo	Sept. 11, 1936 9 yr 2½ mo	May 5, 1937 10 yr 0 mo	10 mo

adolescent sterility is fairly high, for it is unlikely that it is purely due to accidental factors that in these four cases, for which we have reliable evidence, the facts should all point the same way. Furthermore, such evidence as is provided by the cases of Marianne and Boo-Boo, though not altogether satisfactory, negatively supports the existence of such a period of sterility in these particular animals, indicating in the latter case that when a sufficient length of time has elapsed following the appearance of menarche the animal is capable of conceiving promptly when put together with a mature male. In the case of Boo-Boo the conditions parallel those existing among human groups, where marriage and effective intercourse do not occur until a considerable interval after the establishment of menarche, when the promptitude with which conception follows intercourse obscures the existence of an adolescent sterility period in the human female.

It is necessary to be quite clear here concerning the fact that with the exception of Yerkes' data all ages given for the animals referred to above are estimated. Schultz, who was our best authority on the estimation of the age of primates, may be relied upon to have given as nearly accurate an age for Evo as possible. Even so it is not unlikely that the age of this animal was somewhat less or more at menarche than the age estimated by Schultz. Dental age, as Schultz has himself shown, is not a reliable index of chronologic or even of physiologic age.[6] In the case of Dwina, Schultz is of the opinion that she was less than eight years of age at menarche, rather than nine years as estimated by Tinklepaugh.

Marianne was probably two years older than the five years assigned to her by Fox at menarche; while Boo-Boo's estimated age of eight years at menarche probably represents a near approximation to her true age.

Two other chimpanzees for which information is available concerning inception of sexual skin swelling and menstruation may be added to our number here. In the animal "S" at the London Zoological Gardens, Zuckerman noted sexual skin swelling for two years before menarche.[7] In the animal Kani, of the Department of Physiology at Yale University, Zuckerman and Fulton observed the first sexual skin swelling when the animal was at an estimated age of seven years, irregular fluctuations in the amount of swelling being noted during the subsequent years.

Arranging the data for first sexual skin swelling and menarche, as in Table 6, certain interesting facts are seen to emerge.

Since the ages given in Table 6 are, with one exception, all rather hazardous estimates, and the number of cases small, it would be unwise to attempt any elaborate analysis of the findings here presented. On the basis of such facts, however, it is possible to suggest that the average age at the appearance of the first sexual skin in the chimpanzee is 8 ± 1 years, and the age at menarche 8 years and 6 ± 15 months. The sexual skin-menarche interval would thus be some 6 ± 3 months duration. It is of interest to note here that in a paper published in 1937 by Spence and Yerkes on the Florida colony of chimpanzees the authors found that "the average age of seven animals (six of estimated age) at the time of the first menstruation was 8.0 years, the range being 7.25 to 8.33 years."[8] As will be observed, these figures are remarkably close to those which have been given above.

In a valuable paper published in 1943, Young and Yerkes give the exact age of menarche for seven chimpanzees, and the estimated age for an additional 11. In the seven animals in which the exact age was known the mean age at menarche was eight years and 11 months, with a range of seven years four months to 10 years and two months. In the chimpanzees for whom date of birth was unknown, the estimated mean age at menarche was eight years and three months, with a range of seven years to nine years and six months.[9] The mean age of menarche in the chimpanzee may therefore be taken to be eight years and seven months.

Of the 18 subjects reported by Young and Yerkes 13 have been pregnant. Of these, the first pregnancy was normal in seven. "In 2, Lida and Mamo, the first pregnancy

Table 6
The Sexual Skin in Relation to Menarche in the Chimpanzee

Observer	Animal	Age at First Sex Skin	No. Sex Skin Cycles Prior to Menarche	Age at Menarche	Interval Between 1st Sex Skin/ Menarche
Tinklepaugh 1933	Dwina	8 yr 8 mo	4	9 yr	4 mo
Zuckerman and Fulton 1934	Kani	7 yr	8?	Never menstruated	—
Yerkes 1935	Cuba	7 yr 1 mo	3	7 yr 4 mo	3 mo
Schultz and Snyder 1935	Evo	9 yr 1 mo	Continuous swelling 8 mo	9 yr 9 mo	8 mo
Schultz and Snyder 1935	Mae	9 yr 1 mo	1+?	5 weeks later sex swelling reappeared; no menstruation	?
Schultz and Snyder 1935	Pandora	8 yr 6 mo	Continuous swelling 14 mo	No menstruation	14 mo+?
Elder 1938	Soda	7 yr 11 mo	2	8 yr 4½ mo	5½ mo

was terminated by abortion about a month after conception. In Bimba abortion occurred after 42 days. In Bentia, Josie and Mimi, miscarriage occurred the 224th, the 189th, and the 167th day respectively. Three of the 6 subjects (Bentia, Bimba, and Josie), in which the first pregnancy terminated in miscarriage, have since had normal babies. One (Mimi) has had a second miscarriage, 1 (Mamo) was the subject from which a normal appearing 10.5-day embryo, and later a pathological embryo, were removed, and 1 (Lita) had no opportunity for a second pregnancy."[9]

Here, again, it will be observed that the first pregnancy terminated in a miscarriage in six out of the 13 chimpanzees which became pregnant, that is, in 46.1 percent of cases. Again, the evidence points to the relative unreadiness of the adolescent organism for reproduction.

In the seven subjects in which the first pregnancy terminated normally the mean length of the interval between menarche and the first conception is 11.33 months, the range four to 17 months. In the six subjects in which the first pregnancy was terminated by abortion or stillbirth the mean length of the menarche-conception interval was two years, two months, and 22 days, the range being seven months 16 days to five years 28 days.

The difference in the latter group is accounted for by the records of Bentia and Mimi, in which the first fertile matings occurred five years one month, and three years nine and one-half months, respectively, after menarche. There is reason to believe that the prolonged period of sterility in these animals was due to factors other than those associated with adolescence. When these two animals are excluded there are four subjects left in whom the mean menarche-conception interval was 13.66 months. In these four animals pregnancy terminated prematurely. The mean menarche-conception interval for these 11 animals is one year. Young and Yerkes suggest that the mean length of the adolescent sterility interval in the chimpanzee in the Orange Park colony is one to one and one-half years.

Jane van Lawick-Goodall, writing of the nine females

whose adolescence she was able to watch in the Gombe Stream Chimpanzee Reserve in Tanzania, states that "not one has given birth until at least two years after her first adult swelling. . . . Certainly, though, the time lag is beneficial to a wild chimpanzee female: at nine years of age she is not socially mature, nor is she completely full-grown, and she has quite enough to cope with without the added burden and responsibility of a baby."[10]

Returning to Table 6, what is clearly brought out in that table is that there is a considerable amount of variation, in and as between different animals, in respect of the age of appearance of the first sexual skin swelling and in the duration of the interval between the first sexual swelling and the menarche. The same holds true, as Schultz and Snyder have shown, for the menstrual cycle. Such variability is, of course, to be expected, and the differences observed are probably to be ascribed to individual differences in rate or degree of development of the reproductive system; but this is a problem for future research to determine.

The first sexual skin swelling of the chimpanzee is the outward sign of beginning reproductive maturation, the overt sign of the first phase of this process. In man such a premenarcheal phase undoubtedly also occurs, but of this external evidences akin to those seen in the sexual skin of the chimpanzee seem neither to have been sought nor demonstrated. Although the fact is easily confirmed, John Hunter appears to have been the first (ca. 1787) to have drawn attention to the changes in the pudendal skin color with prepubertal growth and also during pregnancy in the human female. He writes: "The skin of the pudenda grows redder and redder to the years of puberty, and seems to have little or no rete mucosum. When a woman is with child these parts grow darker, even darker than the other skin, like the nipple; which colour does not terminate at once, but is gradually lost. This darkness extends to the 'nymphae' and 'carunculae myrtiformes.' "[11]

Such changes, of course, reflect the activity of certain internal secretions. It is probable that the sex hormones poured into the blood stream during the normal estrus cycle of the human female produce changes in the skin of the pudendal region that are related to the sexual skin phenomena of nonhuman primates, although in the human female the well-known menstrual skin changes tend to be rather more diffuse then localized.[12-17] It is, however, quite possible that such changes are in the human female limited to the pubertal, menstrual, pregnancy, and lactation periods. In this connection it may be noted here that while in most of the catarrhine primates in which the sexual skin has been reported the condition appears to be a cyclical phenomenon associated first with beginning ovarian activity and eventually with imminent ovulation, it is, according to Hartman, in the rhesus monkey limited to the period of adolescence, and only rarely appears in the adult.[18,19] This observation has been confirmed by Zuckerman, who has elsewhere also stated that the rhesus monkey rarely experiences *marked* sexual skin swelling.[20,21] In 1938 Schultz showed that the sex skin swelling occurs in the Bornean orang, but this appears to be limited to the period of pregnancy.[22] Sex skin phenomena such as occur in the chimpanzee have never been observed in any of the developing or mature gorillas that have been under observation in various zoological gardens. Raven, however, stated in 1936 that while trailing a group of mountain gorillas (*Gorilla gorilla beringei*) west of the southern end of Lake Kivu in the Republic of Zaire, he saw, at a distance of some 40 yards, an apparently adult female: "The genital region of this individual was swollen, pale in color, and appeared to be tumid. I judged that the swelling was similar in size to that of a female chimpanzee during maximum enlargement, and thus proportionately smaller, considering the larger size of the gorilla."[23] Then the animal disappeared from view. It may be that 40 yards is not a satisfactory distance at which to judge the state of the genital region of a departing gorilla; but anyone who has once seen the sexual skin swelling of a chimpanzee during its maximum enlargement would have very little difficulty in recognizing it at such a distance. Raven was an excellent observer with unrivaled experience in the field; his observation is therefore not to be lightly dismissed.

The late Dr. C. V. Noback, pathologist to the New York Zoological Society, shortly before his death informed me that the postpubertal gorilla (*Gorilla gorilla*), Janet Penserosa, at the New York Zoological Park, which he had had under more or less continuous observation since her infancy, although she had menstruated a number of times never exhibited any circumanal skin changes. However, a premenstrual swelling of the labia majora occurred; that is, before each menstruation, according to Noback, there was a turgid swelling of the labia majora, which ceased during menstruation and did not recur until shortly before the next menstruation. This fact was briefly reported by Noback in 1936.[24] At Dr. Noback's invitation the present writer was able to confirm the presence of such a swelling on two separate occasions (March and October, 1936), but it is not possible to say exactly when these swellings occurred in relation to the estrus cycle. Upon examination the labia majora were found to be distinctly tumid and remarkably soft to the touch, yielding easily upon light pressure. The color was a glistening black, very distinguishable from the dry scaly gray-black color of the rest of the perineal region.[25] Labial swelling reported in the gorilla is clearly related to swelling of the chimpanzee's sexual skin.

The sporadic distribution of the sexual skin among the primates led Zuckerman to suggest that this character has developed independently among the different primates.[26]

In 1886 Hartmann referred to "meist farbigen Zeichnungen der weiblichen Genitalen eines ältern Gorilla-Weibchens vor, welches, obgleich nur schlecht in Weingeist conserviert," but it is not certain whether this may be taken as evidence of the existence of a sexual skin in this animal, which was presumably of the subspecies *gorilla*.[27]

In the animal (*Gorilla gorilla*) Pussi (which was at the Zoological Gardens of Breslau from 3 September 1897, when she was estimated to be four years of age, to the date of her death from chronic nephritis on 6 October 1904) though carefully sought for no evidence of menstruation or of swelling of the labia was ever observed.[28] Zuckerman states that certain anatomical observations of Gerhardt in 1906 seem to support the view that the gorilla shows sexual skin changes.[26] But this must be an error for Gerhardt explicitly states that "periodische Blutungen aus den Genitalen oder Schwellungen der Schämlippen, die sich sonst bei den erwachsenen Schimpansen finden sind hier nie aufgetreten."[29] Nor did Director Grabowsky ever observe such changes. Section of Pussi's ovaries revealed a total absence of ripe ova or any evidences of corpora lutea. Yet this animal, Gerhardt believes, was almost certainly not less than 11 or 12 years old at the time of its death.

The writer has calculated from the published data of Noback that Janet Penserosa was nine years and seven months of age at the time (29 March 1936) of her first observed menstruation, which is an estimate, even though she was suffering from poliomyelitis during this period, probably nearer the true age of menarche in the gorilla than one would be led to believe from the conditions observed in Gerhardt's specimen.[30-32]

Nadler has observed tumescence of the perineal labia in seven lowland gorillas (*Gorilla gorilla gorilla*) of estimated ages from seven to nine and one-half years. It was found that a systematic tumescence of the labia approximated a 31 to 32 day cycle, as compared with a mean of 36 to 37 days in the chimpanzee. Maximal swelling of the labia persisted for an average of 2.2 days in the gorilla as compared with 10 days of sexual skin swelling in the chimpanzee. The swelling of the labia in the gorilla and of the sexual skin in the chimpanzee is highly correlated with the duration of sexual receptivity.[33] Graham and his coworkers have shown that sexual skin swelling in the chimpanzee is produced by an elevation in the level of estrogen. It may be assumed that a similar rise in estrogen in the gorilla is responsible for its cyclic labial tumescence.[34]

In condition of genital swelling and age at menarche, the gorilla would appear to resemble humans more closely than the chimpanzee.

Physiologically and developmentally the sexual skin swelling of the chimpanzee signifies that a gonadotropic hormone elaborated by the anterior lobe of the pituitary sufficient to produce maturation of the follicle in the ovaries, but generally insufficient to produce ovulation, stimulates the release of the follicular estrogenic hormone estradiol* which, among other things, produces an accumulation of intercellular fluid in the perineum, the change thus produced being observable during the period of heightened sexual response as swelling of the sexual skin.[20,21,35-37] It is suggested here that the first appearance of the sexual skin in primates is akin to the rupture of the vaginal plate in mice and rats, for the evidence indicates that the two events are fundamentally due to the operation of similar factors in the two groups of animals. In the guinea pig the period immediately preceding the first estrus, the proestrum, is marked by congestion and swelling of the external genitalia.[38] The same holds true for dogs.

It is important to note here that in the chimpanzee Kani, reported by Zuckerman and Fulton, when, after a year of irregular skin swelling without menstruation having yet occurred, the ovaries were removed and serially sectioned, it was found that none of the follicles was larger than 2 mm in diameter, most being in process of atresia.[28] No luteal tissue was anywhere observed, indicating that follicular growth had never proceeded beyond this stage, and that ovulation had never occurred. Thus, after a year of sexual skin activity, toward the end of which, it may be mentioned, the animal was not in good health, Kani had neither menstruated nor ovulated. It is, of course, very un-

---

*There are three estrogens: estrone, or E, converted from androstenedione or estradiol. Androstenedione is formed mainly by the adrenal cortex and partly in the ovary; estradiol or E_2, the true ovarian estrogen secreted by the ovary; and estriol or E_3, mainly converted in the liver from estrone and more circuitously from estradiol.

likely that ovulation would normally ever occur before menstruation. Ovulation, conception, and parturition may all occur in a female who has never menstruated, but such cases are the exception and must be regarded as abnormal. It may be safely assumed that ovulation as a rule normally takes place at some time after the first estrus or menarche.

Such local changes, rupture of the vaginal plate, edema of the external genitalia or sexual skin swelling, are clearly to be regarded as parallel stages falling into the period of puberty in the development of the reproductive functions; for it is fairly clear that at this stage of sexual development neither mouse, rat, guinea pig, nor chimpanzee is ready for normal reproduction.

Such changes represent the phenomena of the progress of reproductive maturation preceding the development of those processes that result in maturity or the functional capacity to reproduce viably.

We may then, with Harms, recognize three phases* in the adolescent sexual development of the female chimpanzee:

1. A preadolescent phase characterized by sexual skin cycles, without menstruation
2. An anovulatory phase with sexual skin and menstrual cycles
3. An ovulatory phase, which leads into the period of maturity[39]

With respect to the duration of the adolescent sterility interval in the chimpanzee the variability is considerable, from four months to more than two years. This is, in fact, the sort of variability one would expect. The mean length of the sterility interval in the chimpanzee is about one year and three months.

As in all the mammalian forms thus far surveyed, the chimpanzee is, in a certain proportion of cases, capable of

---

*The term "phase" is here used to mean a stage of development. Harms uses the term "period."

conceiving at or shortly after first estrus; and this is probably true of all mammals including humans, but such cases are in the minority, and more often than not represent individuals who are biologically unprepared to carry out the processes of reproduction efficiently. The so-called adolescent sterility period is merely a measure of the chronologic time during which the organism is developing toward the attainment of a condition of *efficient* sexual status, that is, maturity. It is clear that this status is attained only gradually following the establishment of the first overt signs of puberty. Differences in the rates of development of different individuals will, of course, determine when in relation to puberty this status is attained in chronologic time. A perfect demonstration of this fact was provided in 1937 by Engle and his coworkers for first estrus and growth rate in the rat.[40]

In terms of physiologic time it is not difficult to understand that an animal may mature very rapidly after the appearance of the first estrus, and reach a physiological stage of maturity equal to that of an animal that has matured at an appreciably slower rate and is chronologically appreciably older at maturity. In chronologic time there will be a great difference in the duration of the sterility interval; in physiologic time there may be none. This may be stated both as a fact and as an hypothesis to account for the observed differences among different animals in the interval elapsing between first estrus or menarche and the development of efficient reproductive capacities.

Commencing with a number of independent observations on the infertility of postmenarcheal adolescent females in nonliterate societies, it has been seen that such a period of infertility is characteristic of all mammals that have thus far been investigated, and of at least one bird. A period of adolescent sterility is almost certainly a general mammalian trait, and is probably characteristic of most, if not all, animals. We have now to inquire more particularly into the evidence for the existence of such a period of adolescent sterility in humans, and also to discuss the physiological bases of the phenomenon.

REFERENCES

1. Tinklepaugh, O. L. Sex cycles and other cyclic phenomena in a chimpanzee during adolescence, maturity, and pregnancy. *J Morpholog*. 54: 521–47, 1933.

2. Yerkes, R. M. A second-generation captive-born chimpanzee. *Science*. 81: 542–3, 1935.

3. Schultz, A. H., and Snyder, F. F. Observations on reproduction in the chimpanzee. *Bull Johns Hopkins Hospital*. 57: 193–205, 1935.

4. Fox, H. The birth of two anthropoid apes. *J Mammal*. 10: 37–51, 1929.

5. Wyatt, J. M., and Vevers, G. M. On the birth of a chimpanzee recently born in the Society's gardens. *Proc Zool Soc London*, Part I. 195–7, 1935.

6. Schultz, A. H. Eruption and decay of permanent teeth in primates. *Am J Phys Anthropol*. 19: 489–581, 1935.

7. Zuckerman, S. The menstrual cycle of the primates. Part I. General nature and homology. *Proc Zool Soc London*. Series A: 691–754, 1930.

8. Young, W. C., and Yerkes, R. M. Factors influencing the reproductive cycle in the chimpanzee; the period of adolescent sterility and related problems. *Endocrinology*. 33: 121–54, 1943.

9. Spence, K. W., and Yerkes, R. M. Weight, growth and age in chimpanzees. *Am J Phys Anthrop*. 22: 229–46, 1937.

10. van Lawick-Goodall, Jane. *In the Shadow of Man*. Boston: Houghton Mifflin Co., 1971, p. 183.

11. Hunter, J. *Essays and Observations on Natural History, Physiology, Psychology, and Geology*. Edited by Richard Owen. London: Voorst, 1861, 192–3.

12. Bandler, M. I. *Med Gynecol*. Philadelphia: W. B. Saunders, Co., 1927.

13. Hanes, W. J., and Crew, R. S. Endocrinologic and metabolic observations in exfoliative dermatitis. *Endocrinology*. 24: 404–13, 1939.

14. Thorn, G. W., Nelson, K. R., and Thorn, D. W. A study of the mechanism of edema associated with menstruation. *Endocrinology*. 22: 155–63, 1938.

15. Ehrenfest, H. Menstruation and its disorders. *Am J Obstet Gynecol*. 34: 22–4, 1937.

16. Desaux, A. Les rèactions cutanées de la puberté. Edited by G. Laroche. In *La Puberté*. Paris, 1938, 250–73.

17. Garn, S. M., Selby, S., and Crawford, M. R. Skin reflectance during pregnancy. *Am J Obstet Gynecol*. 72: 974–6, 1956.

18. Montagu, A. *An Introduction to Physical Anthropology*. 3rd Edition. Springfield, Ill.: Charles C Thomas, 1960.

19. Hartman, C. G. Studies in the reproduction of the mon-

key *Macacus (Pithecus) rhesus,* with special reference to menstruation and pregnancy. *Carnegie Institution of Washington, Contributions to Embryology,* No. 134, 24, 1932.

20. Zuckerman, S. The physiology of fertility in man and monkey. *Eugenics Rev.* 28: 27–50, 1936.

21. Zuckerman, S. The duration and phases of the menstrual cycle in primates. *Proc Zool Soc London.* Series A: 324, 1937.

22. Schultz, A. H. Genital swelling in the female orang-utan. *J Mammal.* 19: 363–66, 1936.

23. Raven, H. Genital swelling in a female gorilla. *J Mammal.* 17: 416, 1936.

24. Noback, C. V. 1. Note on menstruation in the gorilla (*Gorilla gorilla*). 2. Note on gross changes observed in the external genitalia of the female gorilla just before, during and after menstruation. *Am J Phys Anthropol.* 21 (suppl): 9, 1936.

25. Noback, C. R. Report on his father's findings on the changes in the vaginal smears and associated cyclic phenomena in the lowland gorilla. (*Gorilla gorilla*), *Anat Rec.* 73: 209–25, 1939.

26. Zuckerman, S. *Functional Affinities of Man, Monkeys, and Apes.* New York: Harcourt, Brace & Co., 1933, p. 47.

27. Hartmann, R. Über die Weiblichen Geschlechtstheile der anthropoiden Affen und die Brunst der Affen im Allgemeinen. *Zeitschrift fur Ethnologie.* 18: 431–33, 1886.

28. Grabowsky, F. Beitrag zur Biologie des Gorilla. *Jenaische Zeitschrift für Naturwissenschaft.* 41: 608–11, 1905.

29. Gerhardt, U. Die Morphologie des Urogenitalsystems eines weiblichen Gorilla. *Jenaische Zeitschrift für Naturwissenschaft.* 41: 632–54, 1906.

30. Noback, C. V. Digital epiphyses and carpal bones in the growing infant female gorilla. *Zoologica.* 11: 117–51, 1930.

31. Noback, C. V. The Zoological Park's young female gorilla Janet Penserosa. *Bull NY Zool Soc.* 34: 74–104, 1931.

32. Noback, C. V. 1. Note on menstruation in the gorilla (*Gorilla gorilla*). 2. Note on gross changes observed in the external genitalia of the female gorilla just before, during and after menstruation. *Am J Phys Anthropol.* (suppl) 21: 9, 1936.

33. Nadler, R. D. Cyclicity of tumescence of the perineal labia of the female lowland gorilla. *Anat Rec.* 181: 791–8, 1975.

34. Graham, C. E., Collins, D. C., Robinson, H. et al. Urinary levels of estrogens and pregnandiol and plasma levels of progesterone during the menstrual cycle of the chimpanzee: relationship to the sexual swelling. *Endocrinology.* 91: 13–24, 1972.

35. Zuckerman, S., and Parkes, A. S. The menstrual cycle of the primates. V. The cycle of the baboon. *Proc Zool Soc London.* Series A: 129–91, 1932.

36. Zuckerman, S., and Fulton, J. F. The menstrual cycle of the primates. Part VIII. The oestrin-withdrawal theory of menstruation. IX. The effects of oestrin on denervated skin. *Proc Roy Soc.* Series B. 118: 13–33, 1935.

37. Zuckerman, S., and Fulton, J. F. The Menstrual cycle of the primates. Part VII. The sexual skin of the chimpanzee. *J Anatom.* 69: 1934.

38. Ishi, O. Observations on the sexual cycle of the guinea pig. *Biological Bull.* 38: 237–50, 1920.

39. Harms, J. W. Fortpflanzungsbiologie. Edited by A. H. Schultz, A. Hofer, and D. Starck. In *Primatologia.* 1, Basel & New York: S. Karger, R. G., 1956, 602–3.

40. Engle, E. T., Crafts, R. C., and Zeithamel, C. E. First ertrus in relation to age, weight, and length. *Proc Soc Exper Biol & Med.* 37: 427–32, 1937.

5
ADOLESCENT STERILITY IN HUMANS

Schultz and Snyder remarked in 1935 that "conclusive evidence of adolescent sterility in man is difficult to obtain."[1] Whatever the case may have been in 1935, the evidence available today concerning the existence of such a period of adolescent sterility in the human female is beyond question.

From the historical viewpoint the subject of adolescent sterility is of considerable interest. In the writings of the great Hindu physician Suśruta, who flourished some five centuries before the Christian Era, there occurs the following remarkable passage: "If a man under twenty-five deposit his germ (*garbha*) in a woman younger than sixteen, it will (most likely) die in the womb. Even if it be born alive it will either soon die, or he will be imbecile and weakly so long as he lives."[2]

Plato (340 B.C.?) has this to say in *The Republic*:

"A woman may begin to bear children to the State at

twenty years of age, and continue to bear until forty; a man may begin at five and twenty, when he has passed the point at which the pulse of life beats quickest, and continue to beget children until he be fifty-five.

"Certainly . . . both in men and women those years are the prime of physical as well as of intellectual vigor."[3]

In the *History of Animals* Aristotle writes:

"To begin with, in most animals the secretion of the seminal fluid and its generative capacity are not phenomena simultaneously manifested, but manifested successively. Thus, in all animals, the earliest secretion of sperm is unfruitful, or if it be fruitful the issue is comparatively poor and small. And this phenomenon is especially observable in man, in viviparous quadrupeds, and in birds; for in the case of man and the quadruped the offspring is smaller, and in the case of the bird, the egg.

". . . Man begins to possess seminal fluid about the age of fourteen, and becomes generatively capable at about the age of twenty-one years."[4]

"After the age of twenty-one women are fully ripe for child-bearing, but men go on increasing in vigour."[4]

"At the outset and till the age of one and twenty the spermatic discharge is devoid of fecundity; afterwards it becomes fertile, but young men and women produce undersized and imperfect progeny, as is the case also with the common run of animals. Young women conceive readily, but, having conceived, their labour in childbed is apt to be difficult."[4]

"The beginning of child-bearing in women and the capacity to procreate in men, and the cessation of these functions in both cases, coincide in the one case with the emission of seed and in the other with the discharge of the catamenia: with this qualification that there is a lack of fertility at the commencement of these symptoms."[4]

It is quite clear that 2,300 years before the present Aristotle was fully aware of the existence of a period of adolescent sterility in both humans and animals. He was also aware of the undesirable consequences to mother and child of early pregnancy, as he was aware of the psychological and social unpreparedness of the young adolescent for maternity, as is evidenced by the following passage in the *Politics* in which he wrote:

"Since the legislator should begin by considering how the frames of the children whom he is rearing may be as good as possible, his first care will be about marriage—at what age should his citizens marry? . . . The union of male and female when too young is bad for the procreation of children, in all other animals the offspring of the young are small and ill-developed, and with a tendency to produce female children, and therefore also in man, as is proved by the fact that in those cities in which men and women are accustomed to marry young, the people are small and weak; in childbirth younger women also suffer more, and more of them die; some persons say that this was the meaning of the response once given to the Troezians—the oracle really meant that many died because they married too young; it had nothing to do with the ingathering of the harvest. It also induces to temperance not to marry too soon; for women who marry early are apt to be wanton; and in men too the bodily frame is stunted if they marry while the seed is growing (for there is a time when the growth of the seed, also, ceases, or continues to but a slight extent). Women should marry when they are about eighteen years of age, and man at seven and thirty; then they are in the prime of life."[4]

Matthews Duncan, writing in 1886, suggested that "in the case of some peoples, facts might be collected regarding wives so young as to be in a large proportion sterile from immaturity; and their fecundity gradually appearing as age advanced, might produce a column of mothers from ten to twenty years of age, showing a gradually increasing fecundity of the population at these ages."[5]

Before turning to an examination of the evidence that Duncan himself presented in his distinguished book, we may here note what, after a careful search, proved to be the earliest discoverable modern recognition of the phenomenon of adolescent sterility in man. This is to be attributed to the French physician Renauldin, and occurs in his article "Nubilité" in the great *Dictionaire des Sciences Médicales* of 1819. Renauldin writes:

"NUBILITY, s.f., *nubilitas,* the aptitude for marriage.

"Nubility is not to be confused with puberty; the latter has a slow or rapid beginning and development, which coincides with the last efforts of general growth: the former assumes this growth terminated, and all the organs having reached the degree of perfection and power necessary to permit man to procreate his kind, and woman to bear the fatigues of pregnancy, childbirth, and its aftermaths."[6]

This is beautifully clear. Yet in the same work, Maygrier writing on menstruation equates the first appearance of the latter with the capacity to procreate.[7] It is not until more than a quarter of a century later that we obtain a statement concerning the phenomenon based on detailed observations that are carefully reported. This is by the English physician, J. M. Waddy. In 1846 he carried out his inquiries among the factory hands of the district of Birmingham in England, and the following are his observations on "the age at commencement of menstruation:

"Of 623 females, in one the catamenia occurred at nine years of age; two menstruated at ten, 15 at eleven; 46 at twelve; 87 at thirteen; 130 at fourteen; 115 at fifteen; 105 at sixteen; 67 at seventeen; 43 at eighteen; 10 at nineteen; and two at twenty."[8]

The average age at menarche of the English working-class females was 14.2 years.

Waddy goes on to add that:

"Out of 628 females, we find marriage took place in upwards of 500 persons previous to the twenty-fourth year of their age. Yet few pregnancies took place before that period. The greatest number of pregnancies range from the twenty-fourth to the thirty-fourth year of age. The largest number of marriages are at the nineteenth and twentieth years; and yet only six individuals were pregnant at nineteen years old, and only twenty-seven at twenty. Surely these facts show that the powers of procreation are feeble in the female of tender years; and where Nature indicates a fact, we do well to attend to her instruction."[8]

As far as I am aware Waddy's important observations remained completely unnoticed.

In 1860 the French gynecologist Dubois, in discussing nubility, wrote the following words:

"When this aptitude is developed and when the genital functions may be exercised in woman, without probable prejudice to herself or to her offspring, she is considered nubile.

"Thus understood, nubility not only implies the abstract faculty of procreating, but the possibility of a procreation not injurious to the mother or to the infant.

"Nubility results from the necessary accomplishment of two orders of modification: the one is local and is produced in the organs of generation, the other is general and affects the whole body economy.

"The manifestations of the first, which is also the more important, has for its consequence, puberty, which it is necessary to take care not to confuse, as several authors have done, with nubility, of which it is only one of the essential conditions.

"I wish thus to give their true meaning to two expressions, too often employed as if they had the same signification. The word nubility implies the idea of an aptitude, puberty implies that of a condition which favors or renders possible the exercise of this aptitude. A girl to be nubile must first have puberty. But when she has attained puberty, it does not follow that she is nubile, because puberty is not the only condition necessary for nubility. The ancients made no confusion in this matter, *Puberty,* from *pubes,* hair, indicated the age when certain parts began to be covered with hair. *Nubility,* from *nubis,* a cloud, veil, and from its derivative, *nubilis,* indicated the age when the young girl was in a state to wear the nuptial veil, that is to say, to be married."[9]

Our next observation relating to these matters is from the magnificent gynecological treatise written by Joulin and published in 1867. Joulin writes:

"Nubility is the complement of puberty. These two states should not be confused; they rarely develop at the same time, and their appearance is ordinarily separated by an interval of several years. Puberty is the age at which the young girl becomes a woman; nubility is the period when she becomes capable of all the requirements of maternity. It is frequent, in our latitudes, to see the menses appearing at eleven or twelve years; pregnancy will strictly speaking be possible, but assuming that the young mother has escaped the dangers of a labor so difficult to her, will she be able to nourish her infant, and give it the necessary care? Is she capable of understanding the full extent of her duties and meeting all the difficulties? No, for nubility has not yet arrived, and it is then only that the complete development of the organs, of the bodily powers, and of the intelligence will permit her to fulfill this task adequately. The Civil Code authorizes matrimonial union when the woman is fifteen years old and the man eighteen; but civil law is not in accordance with physiological law, and the race which permits such premature unions will not be long in degenerating. No general limit should be fixed, as is done, for nubility; were this limit fixed at eighteen, twenty years, or twenty-two years, numerous deviations from the common level would occur.

"When I am consulted as to the desirability of marriage for very young individuals, I customarily reply to the parents: that they should not permit their daughter to marry, that is to say, expose her to the risk of becoming a mother until a year at least has elapsed after her stature has ceased to increase. This is the period which I fix upon as nubility; embonpoint, size may add to the volume of the organs, but Nature will add nothing to their development."[10]

A few months later, in 1866/67, we find Matthews Duncan writing:

"It is, I believe, a common notion that the occurrence of menstruation indicates the arrival of the nubile age. Authors occasionally use such expressions as—advent of nubility and commencement of menstruation—as synonymous. The age of puberty may be contemporaneous with the age of nubility; but it cannot be assumed to be so without proof, for very little reflection will suggest to the physiologist many reasons for supposing that the marriageable age is generally delayed for several years after the arrival of the age of puberty."[11]

In corroboration of this point of view Duncan then quotes the passage from Joulin given above. He then goes on to discuss the dangers of early maternity, and quotes two breeders as saying "that the mortality of the young of these animals [lambs and calves], when the mothers are immature, is much greater than when they are well grown. . . . One of them says, 'Taking the first lamb from ewes at one year old has in almost every case failed to be remunerative, owing to the frequent deaths of the lambs. The same may be said of young heifers though the mortality of the offspring may not be so marked as in that of sheep.'[11]

"In conclusion, it is almost useless to add that I consider the age of about from 20 to 25 the nubile age of women. The numerous facts and arguments I have adduced, appear to me to bear out distinctly this conclusion. Below 20 years of age, woman is immature, she runs considerable risk of proving sterile, and if she does bear a child she runs a comparatively high risk of dying in child-bed; besides, her early marriage brings many other disadvantages. . . . The woman above 25 years of age is mature, but to counterbalance this, she encounters some greater risks than the very young wife's though of a similar nature."[11]

In his book, published in 1866, Duncan reprinted the article from which these passages have been taken, together with some additions, as Part VIII (pp. 277 ff.).

Duncan who carried out most of his investigations during the fifties and sixties of the last century, made available a

Table 7
Initial Fecundity of Women of Different Ages Within the First Year of Marriage

Ages Newly Married Wives	15–19	20–24	25–29	30–34	35–39	40–44
No. newly married	700	1,835	1,120	402	205	110
Mothers within 1st yr of marriage	96	339	139	46	19	4
Percentage latter to former	13.71	18.48	12.41	11.44	9.27	3.63

ADAPTED from Duncan, Table 9

large amount of data relating chiefly to Scottish women of the laboring classes, data of the greatest interest in our present connection. In Table 7, for example, the ability of women marrying at different ages to bear children within the first year of married life is clearly described.

From Table 7 it is evident that women of the 15 to 19-year group are less fecund within the first year of marriage than women of the presumed mature group of 20 to 24 years. In the 25 to 29-year group, the ability to conceive within the first year is already less than in the 15 to 19-year group, and steadily declines with age. It must be clearly borne in mind that this table tells us no more than that *the ability to conceive within the first year of marriage* is greatest in the young-mature group, less in the immature group, and increasingly smaller in the older groups. We may legitimately assume that this difference in the first year fertility during the first year in the immature and older groups was due to different factors; in the immature group most probably to incomplete sexual development, and in the older mature groups to aging effects.

When the conditions for the first two years of marriage are considered these startling inequalities are for the most part removed, as is shown in Table 8.

The meaning of this tremendous increase in fecundity in the second year of marriage over the first year of marriage,

Table 8
Initial Fecundity of Women of Different Ages Within the First Two Years of Marriage

Ages of Newly Married Wives	15–19	20–24	25–29	30–34	35–39	40–44
No. newly married	700	1,835	1,120	402	250	110
Mothers within 2 yr of marriage	306	1,661	849	253	84	17
Percentage latter to former	43.71	90.51	75.80	62.93	40.97	15.45
Percent increase over 1st yr	30.00	72.03	63.39	51.49	31.70	11.82

ADAPTED from Duncan, Table 10.

reflected in Table 8, would be difficult to state and foolhardy to suggest here. What is of significance for us here is that in the second year of marriage six times as many women conceived within the 20 to 34-year groups as were able to conceive within the first year of marriage, and more than four times as many in the 35 to 44-year groups; whereas in the relatively immature 15 to 19-year group scarcely more than three times as many were able to conceive in the second year of marriage. Thus, the relative infecundity of the 15 to 19-year group is here interestingly illustrated.

If now we examine the data for the females under 20 years of age, by means of annual instead of quadrennial increments, we shall obtain a clearer view of their initial fecundity during the first and second year of marriage.

From Table 9 we may gather that in the first year of marriage the Scottish females of 16 years of age investigated by Duncan were found to have become mothers in only 6.45 percent of cases; at 17 years of age fecundity is increased by about one-fifth; at 18 years of age almost twice as much as at 17, and two and a quarter times as much as at 16 years of age; and at 19 the increase over the preceding age is one fourth again as great as at 18, and almost three times as great as at 16.

A similar lag in fecundity in the 16-year-old females is

Table 9
Initial Fecundity of Women Under 20 Years of Age in the 1st and 2nd Years of Marriage

Ages of Newly Married Wives	16	17	18	19
No. wives newly married	43	108	225	314
Mothers in 1st year of marriage	2	7	31	56
Corrected percent*	6.45	7.77	14.70	18.30
Mothers within two years of marriage	4	27	98	177
Corrected percent*	12.90	30.00	46.44	57.84

ADAPTED from Duncan, Tables 11 and 12 combined.
* Note: By corrected percent is meant that a correction has been made to allow for the fact that some of the young brides may not yet have commenced menstruating at the time of marriage. This correction tends to lessen the contrast between the fecundity figures of, say, the 16- and 20-year groups. It is possible that this correction was not necessary; it does not seem very likely that females would marry before they had commenced menstruating. At least this would not be the case among Scottish females, upon whom these figures are based.

noticeable in the second year of marriage, for whereas in the 17-year-old brides fecundity has increased by four times over the first year of marriage, 3 1/7 times in the 18-year-old group, and 3½ times in the 19-year-old group, the increase in the 16-year-old group has only been twice as much.

From such evidence we conclude that the Scottish brides reported by Duncan in 1866, at the ages of 16 and 17 years, were significantly less able to conceive within the first year of marriage than those of the 18- and 19-year-old groups. Brides of all four groups showed, in a marked degree, an inability to conceive within the first year of marriage, though this inability was most marked in the youngest group, becoming increasingly less marked with the advance to the 19-year group. Thus, it is seen that the more mature groups exhibit a greater ability to overcome initial infecundity than the younger groups.

It is perfectly legitimate to assume that the conditions of nutrition, occupation, climate, fertility of husbands, methods of intercourse, and contraception were much the same in these several age groups. It would seem very unlikely from Duncan's account that any kind of contraception was practiced. Hence, as far as they go, these data may be accepted as indicative of the fact that immature women, here arbitrarily placed in the 15 to 19-year group, are less frequently able to conceive than mature women, here arbitrarily placed in the 20 to 35-year group.

These conclusions were fully corroborated by the findings on Indian women by Pillai in 1931 in the province of Travancore and by Pearl in 1939 for American women.[12-13]

Duncan's data would indicate that the maximum degree of fecundity is reached at 23 years, before which the capacity is in process of development, and continues until 27 years, whenafter it steadily declines.

Another table from Duncan on a different series of women may be cited here as of interest in our present connection.

Table 10 illustrates once more, in a totally different series of women, the relative infertility of wives under 20 years of age as compared with mature women.

A further table from Duncan illumines the point we have here been discussing.

Initial fecundity is, from Table 11, again seen to be overcome earlier in the mature groups from 20 to 34 than in the immature 15- to 19-year group. The relatively greater initial infecundity of the older group between 35 and 39 may be explained as due to factors quite different from those operating in the case of the immature group; in the latter case we must posit immaturity and all that that implies; in the former case, perhaps the beginning of an involutionary process.

Table 10
Variation in Sterility According to Ages of Wives in Providence

Ages of Brides	Under 20	20–25	25–30	30–40	40–50	Total
Number	144	366	161	108	22	801
First children	49	215	113	45	2	424
Sterile	95	151	48	63	20	377
Percent sterile	65.97	41.25	29.8	58.3	90.9	47.06

ADAPTED from Duncan, Table 71 (2nd ed., 1871, p. 201).

Table 11
The Interval Between Marriage and Birth of First Child in Wives Married at Different Ages

Mothers' Ages at Marriage	15–19	20–24	25–29	30–34	35–39
Number of mothers	649	1905	809	251	96
Average interval from marriage to 1st childbirth (in mo)	18.2	15.9	16.2	18.1	19.1

ADAPTED from Duncan, Table 38 (condensed).

Harris made the following pertinent remarks in the same connection in 1873:

"The *incipiency of menstruation* in our large cities, as a general rule, marks only the gradual approach of the nubile period, and occurs before there are any very decided evidences of womanly development, especially in the immaturity of the pelvic diameters; so that the possibility of conception is still quite remote. Pelvic expansion which appears to have been general in cases of early pregnancy at any age, enabling the subjects of it to bring forth living children of full or nearly full size, is evidently only in its incipiency in a large number of young menstruous girls.

"The *preparatory period,* which usually exists between the first appearance of the menses and the age of possible conception, varies from a few months to several years; but there have been instances in which impregnation followed the first menstrual epoch, or even took place before it has appeared. In tropical countries where young menstrual girls are given in marriage, impregnation very rarely takes place until some time has elapsed, thus marking the duration of this period of sexual preparation."[14]

Mikulicz-Radecki and Kausch, in an investigation carried out in the Woman's Clinic of the University of Berlin Hospital, reported in 1935 that in 74 cases where first sexual intercourse occurred in postmenarcheal girls between the ages of 13 and 17 years inclusive, that only 23 conceived between these same ages, although in approximately 78 percent of these cases no attempt at contraception was made. These investigators considered that this low adolescent fertility is probably due to the fact that ovulation normally follows only some years after menarche. It would certainly appear reasonable to conclude that of the 69 percent of postpubertal females who failed to conceive in this series failure was in the greater number of cases due to the fact that they were physiologically unprepared to do so.

Pearl, in an investigation reported in 1939, relating to the fertility of 956 overtly fertile healthy American women who had been continuously exposed to sexual intercourse, found that fertility within the 15 to 19-year group was very significantly lower than in all the later age groups (from 25 to 39 years inclusive) investigated. These facts are well brought out in Figure 2. Tables 12 and 13 demonstrate changes in median pregnancy rates (per 100 computed ovu-

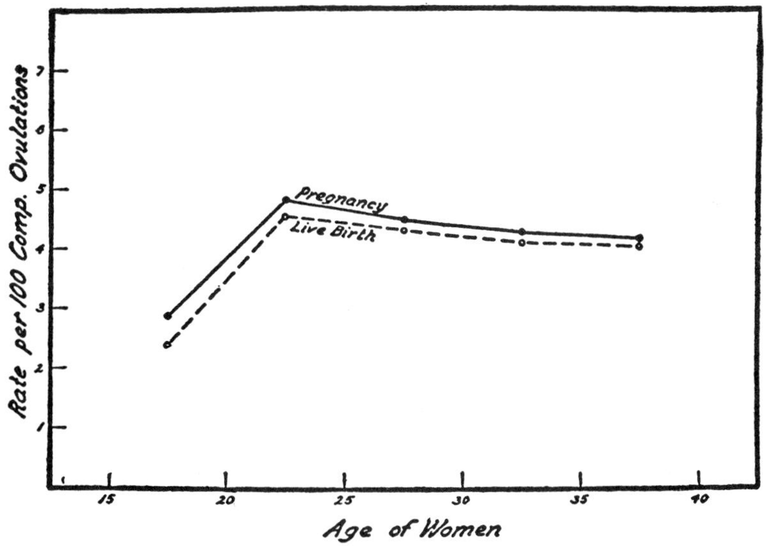

Figure 2. Changes in the unweighted average of median pregnancy and live-birth rates with advancing age in four homogeneous cohorts of women in whose married life no contraceptive efforts were made (Pearl, 1939).[13]

Table 12
Median Fertility Rates of Healthy American Women

Group	N	Median Rates in Indicated Age Periods				
		15–19	20–24	25–29	30–34	35–39
40 years old and over at observation:						
Pregnancy rates	63	2.25 ± 0.59	6.83 ± 0.37	4.78 ± 0.37	4.43 ± 0.44	4.17 ± 0.56
Live-birth rates	63	2.08 ± 0.57	4.90 ± 0.37	4.56 ± 0.30	4.31 ± 0.37	4.03 ± 0.46
35-39 years old at observation:						
Pregnancy rates	169	2.54 ± 0.41	4.88 ± 0.29	....	4.49 ± 0.29	4.21 ± 0.27
Live-birth rates	169	2.27 ± 0.41	4.74 ± 0.25	....	4.38 ± 0.25	4.03 ± 0.26
30-34 years old at observation:						
Pregnancy rates	242	2.75 ± 0.40	4.68 ± 0.23	4.42 ± 0.21	....	....
Live-birth rates	242	2.33 ± 0.39	4.54 ± 0.23	4.20 ± 0.21	....	....
25-29 years old at observation:						
Pregnancy rates	482	3.12 ± 0.40	4.61 ± 0.17	....	....	....
Live-birth rates	482	2.50 ± 0.36	4.42 ± 0.16	....	....	....
Weighted average, all ages at observation:						
Pregnancy rates	956	2.87	4.82	4.49	4.27	4.17
Live-birth rates	956	2.39	4.54	4.31	4.09	4.03

(Pearl, 1939)

lations) with advancing age. Cohorts consist of white multiparae, married once only, living in wedlock, and free of gynecologic disease, married in age group 15 to 19, and thereafter continuously exposed to risk of conception through the age period when they were observed (Table 14). None made any contraceptive effort at any time. Pearl comments upon these findings in the following words:

"The low median fertility in the age period 15–19 . . . may be regarded as the statistical expression of the biological fact, now well established, that in the human species there is a considerable time-lag after puberty before the complete expression of fertility becomes fully established, quite apart from the postponement of marriage."[15]

Some additional evidence from the earlier anthropological literature may now be cited.

Reporting in 1861, the observations of Schurmann, Protector of the aborigines of the Port Lincoln district of South Australia from 1840 to 1846, Wilhelmi, as we have already had occasion to note on an earlier page, states "Although they married so very young (10 to 12 years), the women, according to Mr. Schurmann's observations, generally have no children before the age at which they get them in Europe."[16]

Writing of the Fuegians of the southernmost tip of South America, Bove observed in 1883: "The marriages of the Fuegians are, in general, contracted at an early age; at from ten to twelve the girls are already looking for a husband, but they do not become mothers until they are seventeen or eighteen. The men marry at from fourteen to sixteen years of age."[17]

Since the Fuegians customarily marry immediately after the first menstruation, menarcheal age might be assumed to lie somewhere between ten and twelve years of age among this people—except that European estimates of the age of members of nonliterate cultures are notoriously unreliable. At any rate, Bove seems to have been greatly impressed by the considerable duration of the interval that elapses between marriage and motherhood. His evidence is

Table 13
Standard Deviations of Rates

Group	N	Standard Deviations in Rates at Indicated Ages				
		15–19	20–24	25–29	30–34	35–39
40 years old and over at observation:						
Pregnancy rates	63	5.56 ± 0.33	3.47 ± 0.21	3.47 ± 0.21	4.13 ± 0.25	5.28 ± 0.32
Live-birth rates	63	5.39 ± 0.32	3.48 ± 0.21	2.79 ± 0.16	3.45 ± 0.21	4.30 ± 0.26
35-39 years old at observation:						
Pregnancy rates	169	6.33 ± 0.23	4.39 ± 0.16	4.51 ± 0.17	4.19 ± 0.15	....
Live-birth rates	169	6.24 ± 0.23	3.77 ± 0.14	3.88 ± 0.14	3.94 ± 0.14	....
30-34 years old at observation:						
Pregnancy rates	242	7.38 ± 0.23	4.32 ± 0.13	3.86 ± 0.12	....	....
Live-birth rates	242	7.25 ± 0.22	4.18 ± 0.13	3.80 ± 0.12	....	....
25-29 years old at observation:						
Pregnancy rates	482	10.31 ± 0.22	4.33 ± 0.09	....	....	....
Live-birth rates	482	9.46 ± 0.21	4.13 ± 0.09	....	....	....
Weighted average, all ages at observation:						
Pregnancy rates	956	8.55	4.28	4.04	4.17	5.28
Live-birth rates	956	8.06	4.04	3.69	3.81	4.30

Table 14
Mean and Median Ages at Marriage and at Observation of the Four Cohorts of Women Dealt with Above

Cohort	Age at Marriage		Age at Observation	
	Mean	Median	Mean	Median
First cohort. Observed at 40 and over	18.21 ± 0.10	18.43 ± 0.13	42.48 ± 0.14	41.98 ± 0.17
Second cohort. Observed at 35–39	17.92 ± 0.07	18.04 ± 0.08	37.67 ± 0.06	37.68 ± 0.08
Third cohort. Observed at 30–34	18.15 ± 0.05	18.37 ± 0.07	32.75 ± 0.05	32.70 ± 0.07
Fourth cohort. Observed at 25–29	18.15 ± 0.04	18.39 ± 0.05	27.74 ± 0.04	27.70 ± 0.05

(Pearl, 1939)

not without value as indicating the probable existence of a period of adolescent sterility among Fuegian girls.

In 1901, referring to the Basuto of South Africa, Grützner notes that although intercourse begins at puberty, it is very rare for a girl to become pregnant soon after.[18]

Since effective contraception is not practiced among any nonliterate people of whom we have any knowledge, an observation such as Grützner's, though of doubtful value, is at least worth recording.[19-20]

Allan Webb, Professor of Descriptive and Surgical Anatomy in the Calcutta Medical College, published his observations in 1848 on menarche in Indian girls with the statement, "I believe that even the fact of the existence of this function [i.e., menstruation] having been well established, is no proof of the girl being fit to become a mother, that is, to bear a living child. Almost the only instances I have known here, of instrumental labour in European-bred females, were from their having married too young."[21]

Through the agency of an Indian physician and scholar, Babu Modusuden Gupta, Webb was able to obtain the ages of menarche in 37 Hindu women, 25 of whom had been pregnant and for whom the age at first pregnancy was available. Gupta's data are presented in Table 15.

Gupta's data for the 25 females who had experienced pregnancy show that four of these or 16 percent had conceived within the first year, eight or 32 percent within the third year, four or 16 percent within the fourth year, and one or 4 percent within the sixth year. First pregnancy did not occur until more than a year to three years after menarche in 80 percent of the cases.

The first investigator to collect data systematically at first hand on the reproductive life of a non-European people appears to have been A.-T. Mondière, a physician with anthropological interests who, during the latter part of the last century, spent some six years working among the inhabitants of Cochin China. He reported his observations in 1880, and later, in 1886, reproduced them in summary form in an article. Hartman quoted from this article in his

Table 15
Gupta's Observations on Hindu Women of Calcutta, 1848

No. of Cases	Age at Menarche	Age First Pregnancy	Menarche-Pregnancy Interval
2	11	11	—mo ?
1	11	14	3 yr
7	12	—	—
1	12	12	1 mo ?
1	12	13	1 yr
6	12	14	2 yr
2	12	15	3 yr
5	13	—	—
1	13	13	—mo ?
7	13	14	1 yr
2	13	15	2 yr
1	13	16	3 yr
1	13	18	5 yr
37	Average = 12.3 yr	Average = 14.1 yr	Average = 1 yr 10 mo

1931 communication to *Science* on the relative sterility of the adolescent organism, and in that communication the figures reproduced differ very slightly from those given by Mondière.[22-24] Here Mondière's figures (Table 16) are reproduced from his monograph, which give the essential data in somewhat more detail than in the article from which Hartman quoted.

It is possible that from the figures in Table 16 one year should be deducted from the ages at menarche and birth of first child, to allow for the Chinese method of reckoning age. This would, however, make no alteration in the duration of the sterility interval in the fifth column of figures. Mondière was much interested in the differences presented by the four groups in the duration of the menarche-parturition interval, and suggested that since the climatic conditions were identical for all these groups the difference in the intervals must be attributed to racial factors. He felt, however, that the rather considerable sterility interval in the Cambodian women was possibly in some way connected with the fact that among this people, as a rule, the husband was usually more than twenty years older than his wife. In each of the

Table 16
Mondière's Observations on the Women of Cochin China, 1880

People	Number of Cases	Mean Age at Menarche		Number with First Child	Age at Birth of 1st Child		Sterility Interval	
		years	months		years	months	years	months
Annamite	980	16	4	440	20	6	3	4
Chinese	106	16	6	15	18	10	1	6
Min-Houng	62	16	9	40	20	9	3	2
Cambodian	96	16	10	45	22	6	4	10

NOTE: The calculation of the *sterility interval* from menarche to conception (= −10 months from parturition) has been added by the present writer.

five groups the girl was generally married at the establishment of menarche.

The results of this investigation led Mondière to the conclusion that the arrival of "nubility is more an affair of race than a matter of climate" (p. 824).

Among all these peoples contraceptive practices were entirely unknown and children greatly desired. In a period when "race" served as a convenient explanation for any difference between peoples it is not surprising that Mondière resorted to it as an explanation of the phenomena he observed. In fact "race" seems to have little or no connection with either the onset of puberty or nubility. Whatever Mondière's findings may actually mean, it is unnecessary to do more here than to suggest that they indicate the possible existence of a period of adolescent sterility among the peoples of Cochin China investigated by him.

In his articles "Nubilité" and "Puberté," in the *Dictionnaire des Sciences Anthropologiques,* published between 1886 and 1894, Mondière clearly recognizes the difference between puberty and the period at which actual reproductive capacity appears. In the article on "Nubilité" he writes:

"Nubility is frequently confused with puberty, but it is a very different thing. With the establishment of nubility there follows not only the possibility, but the capacity for normal reproduction. Puberty manifests itself late or early, nubility exists only when the woman is without danger able to give birth to a new being. The sperm in the adolescent, although ejaculated at the period of puberty, does not possess the qualities necessary for producing a regular segmentation of the ovum; and in the young female, if conception takes place, it is doubtful whether the fetus will be brought to term, and the pregnancy is excessively dangerous both for the mother and the offspring. The age of nubility falls into the period of adolescence, that is to say from eleven to twelve years in the female until twenty or twenty-one years, and strictly speaking the term should be applied only to this period'' (p. 823).

Similar views are expressed by Mondière in his article ''Puberté,'' in the same work (p. 941).

In the early 1920s Professor J. Preston Maxwell of Peiping Union Medical College conducted an investigation on 99 married women of Yung Chun in South Fukien, China, relating to menarche, marriage, first delivery and footbinding. These data were never published by Professor Maxwell, and I am greatly indebted to him for his gift of this material and his permission to use it in the present volume (Table 17).

The average age of menarche of these women was 16.13 years, with a range of 13 to 21 years. First effective intercourse took place at an average of 17.38 years with a range of 13 to 23 years, and first childbirth at 20.02 years, with a range of 15 to 26 years. Table 17 represents the distribution in this group of women of the intervals between menarche and first childbirth at various ages. The figures make it clear that in only a small percentage of cases is menarche followed by pregnancy within the first year. Of these women 34

Table 17
Interval Between Menarche and First Childbirth in 99 Chinese Married Women From South Fukien

Interval	Less than 1 yr.	1 Yr	2 Yr	3 Yr	4 Yr	5 Yr	6 Yr	7 Yr	8 Yr	9 Yr	No Pregnancy
No cases	3	6	17	11	15	7	8	5	3	4	20
Percent	3.0	6.0	17.0	11.0	15.0	7.0	8.0	5.0	3.0	4.0	20.0

Table 18
Interval Between Menarche and First Childbirth in 34 Chinese Women Married Before Menarche

Age at Menarche	Number	Age at Marriage	No	Age at 1st Child	Menarche-Childbirth Interval
14–15	14	15–16	11	18–19	3 yr
16	13	17–18	13	19–20	2 yr
16	7	21–23	6	21–23	2 yr 6 mo

had arrived at menarche from one to five years before marriage, as shown in Table 18.

The small number of cases available hardly renders Table 18 an impressive one, and also since Professor Maxwell's data were gathered in terms of years without record of months, this table is probably far from representing the actual facts. However, taking it at its face value it does exhibit something of a general trend, the younger 14 to 15-year group apparently taking a longer time to overcome their initial infecundity, and being more often sterile (three cases) than is the case in the older groups. This is again borne out by the figures for the whole group arranged by age at menarche, as is seen in Table 19.

This table gives a striking illustration of the decrease in the sterility interval as the adolescent female arrives later at menarche. Similarly, Mills and Ogle note that "the sterile lag period from menarche to first conception . . . exhibits a steady shortening as we go from earlier to later menarchial age."[25] Such facts would suggest that the more mature the soma the shorter will be the sterility interval following menarche.

Chau and Wright, in a report published in 1925 of an investigation carried out on 2,291 South Chinese women living in subtropical Canton (N. lat., 23° 7′ 10″) and surrounding districts, found the mean age at menarche to be 14.5 years; the mean age at marriage (683 cases) 17.6 years, and the mean age at birth (596 cases) of the first child 20.5 years. Only one girl became a mother at the age of 13 years; five at the age of 15, and twelve at the age of 16.[26]

Table 19
Menarche-Conception Interval in 99 South Chinese Married Women

No cases	Age at Menarche	Age at Marriage	Age at 1st Child	No of Conceptions	% of Conceptions	Menarche-Conception Interval
6	13	18	21	3	50.0	7
11	14	17	20	2	18.2	5
29	15	16	19	2	7.0	3
16	16	16	20	4	25.0	3
18	17	16	20	3	16.6	2
9	18	18	21	1	11.1	2
3	19	17	21	0	—	1
5	20	17	22	4	40.0	1 } 1 yr
2	21	22	27	1	50.0	5 } 1 yr

Clearly such figures are of very little worth for our purposes, although in a very general way they may be regarded as not without some suggestive value. Were we to accept these data at their face value we should have to suppose an average sterility interval of some 5.2 years. One cannot be certain that such an average sterility interval is at all unlikely. It may be recalled here that among the natives of Wogeo, in New Guinea Territory, menarche generally arrived at seventeen, and despite frequent intercourse during the interim it was not usual for a girl to give birth to offspring until she was twenty-two; this yields a sterility interval of five years. However, until further investigations have been carried out in this and in other regions of the world it will not be possible to say with any degree of certainty whether such an *average* interval occurs in any human group.

Mills and Ogle, in a paper already referred to, presented the following data, collected from hospital records in Hong Kong, Panama, and the United States (Table 20).[25]

From Table 20 it will be noted that the fertility (actually *sterility*) interval appears to be about three years longer in the married than in the unmarried except in the case of the Manila groups, where the interval is practically the same in the married and the unmarried. The authors think the latter coincidence in age at first childbirth (6.33 years after menarche) reflects a real expression of physiologic sterility, since "sex relationships in Manila are perhaps as free from social or moral restraint as anywhere on earth" (p. 611). This may be so, but we are not told when effective intercourse begins in any of these groups, hence it does not seem to me that the figures in the last column can have any real value for our purposes. This is, however, not the case when we examine the material for the interval between marriage and conception. For Cincinnati white and black women the interval is 0.59 years, and for Richmond married blacks it is 0.37 years. In Panama married blacks' conception follows immediately after marriage, while in the married Filipinos at Manila the interval lasted 0.59 years, and some 2.00 years in married Chinese at Hong Kong.

Table 20
Menarche, Marriage, and First Birth Data

City	Subjects	Date	No	Mother's Age in Years at: Menarche	Marriage	1st Delivery	Lag in Fertility (Menarche to first conception)
Cincinnati	Unmarried black	1924	33	13.44 ± 0.13	....	18.08 ± 0.28	3.89
	Married white-black		131	13.68 ± 0.08	19.09 ± 0.17	20.48 ± 0.19	6.05
Richmond, Va.	Unmarried black	1934	56	13.46 ± 0.09	....	18.21 ± 0.19	4.00
	Married black		52	13.75 ± 0.17	19.63 ± 0.17	21.60 ± 0.43	7.10
Panama	Unmarried black	1934	72	14.04 ± 0.09	....	19.26 ± 0.20	4.47
	Married black		56	14.05 ± 0.14	21.01 ± 0.27	21.93 ± 0.32	7.13
Manila	Unmarried Filipinos	1934	22	14.73 ± 0.16	....	21.82 ± 0.50	6.33
	Married Filipinos		84	14.58 ± 0.12	20.49 ± 0.30	21.88 ± 0.30	6.55
Hong-Kong	Married Chinese	1934	284	15.82 ± 0.06	18.27 ± 0.11	21.00 ± 0.12	4.97

ADAPTED from Mills and Ogle, Table 1 (1936).

Hence, the only evidence of any significance yielded by these data is that although Hong Kong Chinese married women arrive at menarche at an average age of 15.82 years, and marry at an average age of 18.27 years, marriage is followed by an average lag before the first conception (as determined by the deduction of 10 months from age at first delivery) of 2.00 years. It seems likely that the greater "fertility lag" in the married as compared with the unmarried is due to the fact that the married generally begin effective intercourse appreciably later than the unmarried. The figures for Manila relating to childbirth in the unmarried and the married may, as Mills and Ogle believe, reflect the actual existence of a physiologic sterility interval here of 6.33 years; we should, however, want to see much supporting evidence before we could accept such a figure as a fact. We have seen that in two Chinese groups such a period of sterility has already been recorded by two independent investigators, but the same criticisms apply to their findings as to those of Mills and Ogle. Incidentally, it is necessary to point out here that Mills and Ogle's last column in Table 20 is incorrectly labeled "lag in fertility." This column obviously reflects nothing of the sort; what it gives is what is stated in the introduced parenthesis, namely, the interval between menarche and conception—a very different thing.

The 2.00 years interval following marriage (2.65 years after menarche) before the first conception among Hong Kong Chinese women may reflect the existence of a period of adolescent sterility, but what this may be due to is another matter. In spite of the statement of Mills and Ogle to the contrary Professor J. Preston Maxwell has informed me that the desire for children has been by no means as great as it formerly was in such places as Hong Kong, and that effective contraceptive practices are far from unknown among the general population. However, the low economic level of subsistence of the Chinese here reported may be an additive factor in complicating the evidence. Such objections, of course, apply to all the evidence that has thus far been cited. The evidence of Mills and Ogle, though clearly

Table 21
Menarcheal Ages and Maternal Ages at First Birth in Panama City

Subjects	No of Cases	Mother's Age in Years at: Menarche	Mother's Age in Years at: First Birth	Lag in Fertility (menarche to first conception)
Married	18	14.22 ± 0.25	20.22 ± 0.72	5.2
Unmarried	132	14.05 ± 0.8	20.08 ± 0.19	5.2
Unmarried girls of				
Cincinnati	33	13.44 ± 0.13	18.08 ± 0.28	3.9
Richmond	56	13.46 ± 0.09	18.21 ± 0.19	4.0
Filipinos in Manila	22	14.73 ± 0.16	21.72 ± 0.50	6.3

NOTE: Mills Table 3, p. 10.

by no means conclusive, does at any rate point in the direction of a probable adolescent sterility interval among the people cited.

In 1942 Mills recorded the results of an investigation of the menarche–first-birth records of 18 married and 132 unmarried Panama City mothers "in whom sexual restraint is probably at a minimum."[27] It will be seen from Table 21 that maternal age at first birth is practically the same whether the mother was married or unmarried. Similar data previously obtained by Mills at Cincinnati, Richmond, and Manila are included in this table for comparison. From this table it will be seen that the menarche-conception interval in these 150 Panama City women was 5.2 years.

A certain amount of material exists for Indian women that may profitably be discussed at this juncture. In a book that enjoyed a wide circulation, *Mother India* by Katherine Mayo, published in 1927, the author drew attention to what she esteemed to be the fact that "the Indian girl, in common practice, looks for motherhood nine months after reaching puberty, or anywhere between the ages of fourteen and eight. The latter age is extreme, though in some sections not exceptional; the former is well above the average."[28]

At the time of their publication these statements, among others, excited much indignation in informed quarters, a feeling that found expression in the Indian press and elsewhere. But long before the advent of Mrs. Mayo's controversial book the child-mothers of the East had become an established superstition of orthodox Western folklore relating to the East. Child mothers there are in India, as there are elsewhere in the world, but they are not usually to be found "between the ages of fourteen and eight."

As long ago as 1845 Robertson showed, on the basis of data collected by W. H. Goodeve, Professor of Midwifery at the Calcutta Medical College, that the average age at menarche of 90 Hindu women (Bengal) was 12 years and four months.[29] In a supplement to his paper, published in the same year, Robertson gave the figures supplied him by Webb, which the latter published in his book of 1848.[30] In that paper and in his book Webb states, "It is not common for girls in India to begin to menstruate until after the twelfth year. . . . Out of a list of 127 Hindoo females with which I have been favored, it began only in six girls under 12 years of age, and as many of them did not again menstruate until a year after this which they believe a *first appearance,* it is probable . . . that a ruptured hymen would account for that. Thus 81 out of 127 are stated to have been 12 years old or upwards.

"Out of eighty cases thus furnished who had probably been subjected to the influences of impregnation from the age of nine years, there were only 28 births under 14 years of age, but similar results would perhaps have followed in Europe . . . out of 127 cases reported to me of Bengalese, one birth is stated at eight, and one at nine."[31]

In 1846 Robertson published a detailed table of the reproductive history of 230 Hindu women representing the data collected by A. H. Leith of the Bombay Medical Service.[32] The average age of these Hindu women was 13 years, five months, and 24 days; the average age at the birth of the first child was 16 years, 10 months, and 22 days. Since intercourse with a mature male had been practiced from the commencement of menarche, and often before, the average sterility interval in this group of 230 women was two years, seven months, and 28 days (Table 22).

Table 22
Cases Collected at Bombay in 1845

BY A. H. LEITH

No.	Caste	Age at menarche			Age at birth of first child			No.	Caste	Age at menarche			Age at birth of first child		
		Y	M	D	Y	M	D			Y	M	D	Y	M	D
1	Brahman	12	—	—	—	—	—	21	Brahman	13	2	—	—	—	—
2	Brahman	12	—	—	—	—	—	22	Brahman	15	1	—	16	9	—
3	Brahman	13	—	—	—	—	—	23	Brahman	18	—	—	—	—	—
4	Brahman	13	8	—	—	—	—	24	Purbhoo	13	6	—	—	—	—
5	Brahman	14	3	—	—	—	—	25	Purbhoo	13	—	27	—	—	—
6	Brahman	11	4	17	—	—	—	26	Purbhoo	12	2	—	—	—	—
7	Brahman	11	1	7	16	3	22	27	Purbhoo	12	—	22	13	4	27
8	Brahman	11	9	—	17	2	15	28	Purbhoo	11	6	—	12	6	—
9	Brahman	12	—	—	19	3	—	29	Purbhoo	11	—	—	—	—	—
10	Brahman	13	3	—	—	—	—	30	Purbhoo	14	—	11	—	—	—
11	Brahman	13	6	5	14	2	19	31	Purbhoo	13	2	7	—	—	—
12	Brahman	12	7	—	—	—	—	32	Purbhoo	13	—	—	—	—	—
13	Brahman	13	—	—	15	—	28	33	Purbhoo	14	7	—	16	2	9
14	Brahman	14	1	—	16	—	17	34	Purbhoo	12	—	—	—	—	—
15	Brahman	14	—	—	15	—	21	35	Purbhoo	14	—	—	—	—	—
16	Brahman	12	6	—	13	6	—	36	Purbhoo	11	5	—	—	—	—
17	Brahman	13	2	—	—	—	—	37	Purbhoo	13	—	—	—	—	—
18	Brahman	12	5	—	13	9	—	38	Purbhoo	15	—	—	—	—	—
19	Brahman	14	—	—	—	—	—	39	Purbhoo	14	—	—	15	—	—
20	Brahman	16	—	—	17	3	—	40	Purbhoo	12	—	—	—	—	—

Table 22 (Continued)
Cases Collected at Bombay in 1845

By A. H. Leith

No.	Caste	Age at menarche			Age at birth of first child			No.	Caste	Age at menarche			Age at birth of first child		
		Y	M	D	Y	M	D			Y	M	D	Y	M	D
41	Purbhoo	15	—	—	17	3	—	61	Bunya	14	—	—	15	—	—
42	Purbhoo	13	6	—	—	—	—	62	Bunya	12	4	—	16	—	—
43	Purbhoo	12	3	—	14	—	—	63	Bunya	13	—	—	—	—	—
44	Purbhoo	16	2	—	—	—	—	64	Bunya	12	—	—	—	—	—
45	Purbhoo	12	2	14	14	8	—	65	Bunya	15	—	—	20	—	—
46	Purbhoo	17	—	—	—	—	—	66	Bunya	12	1	—	—	—	—
47	Bunya	11	—	—	—	—	—	67	Bunya	11	7	—	15	—	—
48	Bunya	12	—	—	—	—	—	68	Bunya	13	10	—	—	—	—
49	Bunya	14	—	—	—	—	—	69	Bunya	15	4	—	—	—	—
50	Bunya	13	—	—	—	—	—	70	Bunya	10	—	—	13	—	—
51	Bunya	16	—	—	—	—	—	71	Bunya	13	—	—	15	—	—
52	Bunya	12	9	—	—	—	—	72	Bunya	12	—	—	15	—	—
53	Bunya	11	—	9	12	6	—	73	Bunya	11	—	—	—	—	—
54	Bunya	11	3	—	—	—	—	74	Bunya	11	—	—	22	—	—
55	Bunya	12	—	—	—	—	—	75	Bunya	10	—	—	—	—	—
56	Bunya	12	—	—	13	9	—	76	Bunya	16	—	—	22	—	—
57	Bunya	14	11	—	15	7	—	77	Bunya	13	—	—	—	—	—
58	Bunya	13	2	—	—	—	—	78	Bunya	12	—	—	13	—	—
59	Bunya	15	1	9	17	1	—	79	Bunya	13	—	—	—	—	—
60	Bunya	17	—	3	—	—	—	80	Bhatia	13	—	—	15	1	—

Table 22 (Continued)
Cases Collected at Bombay in 1845

By A. H. Leith

No.	Caste	Age at menarche Y	M	D	Age at birth of first child Y	M	D	No.	Caste	Age at menarche Y	M	D	Age at birth of first child Y	M	D
81	Bhatia	12	3	—	—	—	—	101	Sonar	11	4	—	—	—	—
82	Bhatia	11	6	—	—	—	—	102	Sonar	13	—	—	—	—	—
83	Bhatia	12	—	8	—	—	—	103	Sonar	13	6	—	—	—	—
84	Bhatia	12	—	—	13	7	—	104	Sonar	15	—	—	—	—	—
85	Bhatia	13	9	—	—	—	—	105	Sonar	13	—	—	14	6	—
86	Bhatia	14	—	—	15	9	—	106	Sonar	12	2	—	—	—	—
87	Bhatia	12	—	—	—	—	—	107	Sonar	15	—	—	16	7	—
88	Bhatia	13	2	5	—	—	—	108	Sonar	14	—	—	—	—	—
89	Bhatia	12	—	—	—	—	—	109	Sonar	12	—	—	14	—	—
90	Sonar	11	—	9	—	—	—	110	Sonar	11	8	—	—	—	—
91	Sonar	11	9	—	14	7	—	111	Sonar	15	—	—	20	—	—
92	Sonar	14	—	—	18	4	—	112	Sonar	13	15	—	—	—	—
93	Sonar	17	—	—	21	3	—	113	Kansur	11	9	25	15	9	—
94	Sonar	15	3	—	—	—	—	114	Kansur	11	—	—	—	—	—
95	Sonar	13	—	—	—	—	—	115	Kansur	14	—	—	—	—	—
96	Sonar	13	—	—	—	—	—	116	Kansur	13	—	—	16	2	—
97	Sonar	11	—	17	—	—	—	117	Kansur	12	—	—	13	6	—
98	Sonar	13	1	—	15	6	—	118	Kansur	13	2	—	—	—	—
99	Sonar	12	3	—	—	—	—	119	Kansur	12	1	9	—	—	—
100	Sonar	12	—	—	—	—	—	120	Kansur	12	—	—	—	—	—

Table 22 (Continued)
Cases Collected at Bombay in 1845

By A. H. Leith

No.	Caste	Age at menarche			Age at birth of first child			No.	Caste	Age at menarche			Age at birth of first child		
		Y	M	D	Y	M	D			Y	M	D	Y	M	D
121	Kansur	11	—	29	—	—	—	141	Simpee	14	—	—	—	—	—
122	Kansur	15	—	—	—	—	—	142	Simpee	14	7	—	17	4	—
123	Kansur	14	—	—	—	—	—	143	Khutree	12	3	—	13	6	—
124	Kansur	12	3	—	—	—	—	144	Khutree	13	—	17	—	—	—
125	Kansur	12	9	—	—	—	—	145	Khutree	17	2	—	—	—	—
126	Kansur	13	—	—	—	—	—	146	Khutree	13	7	—	—	—	—
127	Kansur	13	—	—	—	—	—	147	Khutree	14	2	25	16	3	—
128	Kansur	16	7	16	—	—	—	148	Khutree	19	—	—	24	6	—
129	Kansur	12	5	29	14	1	1	149	Khutree	18	—	—	21	—	—
130	Simpee	14	2	15	15	7	—	150	Khutree	21	—	—	26	4	—
131	Simpee	16	—	—	17	—	1	151	Khutree	17	—	3	—	—	—
132	Simpee	14	—	—	15	6	—	152	Khutree	23	—	—	—	—	—
133	Simpee	12	—	—	13	3	—	153	Khutree	12	3	—	15	6	—
134	Simpee	12	7	—	—	—	—	154	Khutree	14	6	—	16	2	—
135	Simpee	13	—	—	—	—	—	155	Khutree	13	—	—	—	—	—
136	Simpee	13	6	—	—	—	—	156	Khutree	12	6	—	14	6	—
137	Simpee	14	—	—	—	—	—	157	Khutree	14	—	—	—	—	—
138	Simpee	12	—	—	—	—	—	158	Khutree	13	4	—	15	—	—
139	Simpee	15	—	—	19	4	—	159	Khutree	15	—	—	—	—	—
140	Simpee	13	—	—	22	—	—	160	Khutree	11	9	—	14	—	—

Table 22 (Continued)
Cases Collected at Bombay in 1845

BY A. H. LEITH

No.	Caste	Age at menarche Y	M	D	Age at birth of first child Y	M	D	No.	Caste	Age at menarche Y	M	D	Age at birth of first child Y	M	D
161	Sootar	12	2	11	—	—	—	181	Bhunsalee	14	—	—	—	—	—
162	Sootar	12	—	27	—	—	—	182	Bhunsalee	16	—	—	—	—	—
163	Sootar	13	3	—	—	—	—	183	Bhunsalee	11	3	13	15	—	—
164	Sootar	11	7	—	—	—	—	184	Bhunsalee	15	2	21	17	—	—
165	Sootar	11	11	19	—	—	—	185	Bhunsalee	12	3	—	—	—	—
166	Sootar	15	—	—	19	—	—	186	Bhunsalee	17	—	21	19	—	—
167	Sootar	14	—	—	16	—	—	187	Bhunsalee	13	3	22	—	—	—
168	Sootar	17	—	—	19	2	—	188	Bhunsalee	19	3	17	25	—	—
169	Sootar	16	2	—	18	1	20	189	Bhunsalee	21	—	—	—	—	—
170	Sootar	13	3	21	14	1	—	190	Bhunsalee	15	—	—	—	—	—
171	Bhunsalee	14	—	13	16	—	—	191	Muratha	14	—	—	—	—	—
172	Bhunsalee	17	—	19	—	—	—	192	Muratha	12	—	—	—	—	—
173	Bhunsalee	13	3	—	15	2	—	193	Muratha	15	—	—	—	—	—
174	Bhunsalee	11	9	23	—	—	—	194	Muratha	11	4	—	—	—	—
175	Bhunsalee	13	—	—	15	3	—	195	Muratha	12	—	—	—	—	—
176	Bhunsalee	14	—	—	19	—	—	196	Muratha	13	—	—	—	—	—
177	Bhunsalee	15	—	—	19	—	—	197	Muratha	12	—	—	15	—	—
178	Bhunsalee	14	—	—	—	—	—	198	Muratha	15	—	—	20	—	—
179	Bhunsalee	15	—	—	17	3	—	199	Muratha	14	—	—	—	—	—
180	Bhunsalee	13	2	9	—	—	—	200	Muratha	12	—	—	17	—	—

Table 22 (Continued)
Cases Collected at Bombay in 1845

By A. H. Leith

No.	Caste	Age at menarche Y	M	D	Age at birth of first child Y	M	D	No.	Caste	Age at menarche Y	M	D	Age at birth of first child Y	M	D
201	Muratha	17	—	—	22	—	—	216	Parsee	14	2	—	21	—	28
202	Muratha	19	—	—	25	—	—	217	Parsee	12	3	—	25	3	28
203	Muratha	13	3	—	—	—	—	218	Parsee	14	6	15	23	6	—
204	Muratha	12	10	—	16	9	—	219	Parsee	17	—	22	—	—	—
205	Bunduree	14	7	—	21	8	—	220	Parsee	12	6	11	—	—	—
206	Bunduree	13	9	—	16	—	—	221	Parsee	12	2	—	—	—	—
207	Bunduree	16	5	29	19	10	—	222	Parsee	11	1	3	—	—	—
208	Mungalee	12	6	17	14	3	13	223	Parsee	13	3	7	15	—	—
209	Mungalee	15	—	8	18	11	21	224	Parsee	11	—	9	—	—	—
210	Parsee	12	—	—	14	6	21	225	Parsee	11	9	22	13	11	—
211	Parsee	12	2	—	16	9	—	226	Parsee	12	6	21	—	—	—
212	Parsee	13	—	—	—	—	—	227	Parsee	11	2	—	—	—	—
213	Parsee	11	8	17	—	—	—	228	Parsee	14	—	—	—	—	—
214	Parsee	11	5	—	—	—	—	229	Parsee	14	—	—	18	4	—
215	Parsee	13	—	—	18	7	11	230	Parsee	13	8	—	15	4	—

According to the more modern evidence, for example, the Indian Census reports for 1931 "the average of females at marriage is 13.33 years and of males 17.98 years," the females seldom going to live with the husband until after puberty (in the Punjab, at least).[33] From the Indian Census Report for 1921 we learn that the first child in the great majority of cases is born in the third year of effective marriage, that is, dating from menarche.[34] This puts the mean age of the mother at 16 years and some months.

Data relating to the figures from the Maternity Hospital at Seva Sada in Amednagar, in which Presidency early marriage is said to be more prevalent than in any other part of India, indicate that the average age of the mother at first delivery is 18.3 years.[35]

In 1927 Dr. M. I. Balfour of Bombay, in an account of 6,580 primiparous mothers from many parts of India, stated that none was under 13 years of age, seven were recorded as 13, and 35 as under 15. The average age of the mother at birth of her first child was 18.7 years in Bombay, and 19.4 years in Madras. In both Bombay and Madras respectively 85.6 and 86.2 percent of the mothers were more than 17 years of age. These figures show that less than 1 percent of Hindu women were mothers before they were 15 years of age, and none before the age of 13.[36]

An analysis of Curjel's report of 1920 on menstruation in 545 Calcutta women of various social classes yields a mean age of 14.21 years for age at menarche, a figure that would yield an adolescent sterility interval for the Indian woman of approximately four years.[37] It may be noted here that in 1930 Pillay stated that with respect to Calcutta women "an analysis of about 700 cases shows that the age of puberty is 13 or 14, and the age of birth of the first child is 14 or 15."[38]

These statements would, upon the face of them, appear to be extremely doubtful; at any rate, they are lacking in confirmation and, unfortunately, Pillay does not present the detailed figures upon which they are based.

Sen, in 1950, found that of the 216 high caste Calcutta women who married before menarche the mean age of the mother at birth of the first child was 16.25 years.[39] Since the mean menarcheal age was 12.78 years, there was an interval of 3.47 years between menarche and birth of the first child.

The findings that have since become available bearing on the age of menarche in Indian women, as may be seen from Table 23, further serves to underscore the erroneousness of Mrs. Mayo's judgments. Mrs. Mayo herself stated that the youngest mother she saw was 12 years of age, but the probabilities are high that even that mother was older. For it is a fact well enough known by anthropologists and others who have had some experience among non-European peoples, that white observers almost invariably underestimate, by years, the ages of non-European children since children, owing to malnutrition and a variety of other conditions, frequently appear somewhat less developed than white children at similar ages. Reche has suggested that this is in part due to the lateness with which the secondary sex characters appear in such children.[40] This may well be so since it is known that malnutrition affects the development of the secondary sex characters. It is of interest to note here that the appearance of adult members of many non-European peoples is such as to cause white observers almost invariably greatly to overestimate their age.[41] Presumably the same factor, malnutrition, is responsible for causing the children to appear younger and the adults to look older than they really are.

Professor C. A. Mills informed me that during his widely extended investigations among non-European peoples, in various parts of the world, it was invariably his experience that when he was able to check his own estimate of the age of an adolescent or a child by some official birth record, he found his own estimate to be too low. Actually it is very doubtful whether there exists any people among whom the mean age at menarche is less than 12½ years. The most reliable data on age at menarche for different human

Table 23
Age at Menarche in Various Populations

Population	Investigators	Number	Age at Menarche	Range
Africa,				
West, Ashanti	Fortes, 1954	8,975	15.5	
East, Kampala	Burgess et al., 1960	528	13.4 ± 0.16	
East, Baganda	Burgess & Burgess, 1965	168	13.4	
East, Buhaya	Richards & Reining, 1954	303	13.6	
East, Rwanda	Heints, 1963		16.5	
East, Hutu	Hiernaux, 1965		17.13 ± 0.30	
East, Tutsi	Hiernaux, 1965		16.5	
East, Nyakyusa	Hautvast, 1971	110	14.93	
East, Kisi	Hautvast, 1971		15.5	
South, Bantu	Oettle & Higginson, 1961	1,002	14.89 ± 1.66	
South, Bantu	Kark, 1957	528	13.60 ± 1.13	
South, Bantu (poor)	Burrell et al., 1961	17,036	15.42 ± .04	
South, Bantu (non-poor)	Burrell et al., 1961	20,834	15.02 ± 0.03	
South, Bantu (poor)	Frere, 1971	1,688	14.76 ± 0.03	
South, Bantu (non-poor)	Frere, 1971	3,150	14.75 ± 0.39	
South, Coloured, Johannesburg	Margolius, 1970	1,693	13.2 ± 0.42	10–16
South, Kung, Botswana	Howell, 1976	51	16.6 ± 1.3	
South, Whites	Van Castricium, 1946	2,191	14.65 ± 1.79	9–21
South, Whites	Benjamin, 1960	1,000	14.6	
Nigeria, Lagos	Ellis, 1950	542	14.40 ± 1.0	11–19
Ibo	Tanner & O'Keefe, 1960	346	14.07 ± 0.16	11.2–17
Northern (Pagans)	Wilson & Sutherland, 1953	74	14.50 ± 1.24	
Northern (School)	Wilson & Sutherland, 1953	172	14.14 ± 0.04	
Ibadan (Urban)	Oduntun, Ayeni, & Kale, 1976	2,029	13.70 ± 0.03	

Igbo-Ora (Rural)	Oduntun, Ayeni, & Kale, 1976	328	14.50 ± 0.09	
Somalia, Mogadishu	Gallo, 1975	524	14.78 ± 0.07	
America, N, NE, NW	Kinsey, 1953	5,770	13.1 ± 1.2	9–18
Berkeley, Calif.	Nicolson & Hanley, 1953	91	12.8 ± 1.1	10.5–15.5
Berkeley, Calif.	Livson & McNeill, 1962	43	12.8 ± 1.1	
Boston, Mass. (Nursing students)	Zacharias & Wurtman, 1964	26	12.46 ± 0.85	
Boston, Mass.	Damon et al., 1969	78	12.8	
Cambridge, Mass.	Shuttleworth, 1937	315	13.0	
Cambridge, Mass.	Shuttleworth, 1937	248	13.1 ± 1.1	
Newton, Mass.	Zacharias et al., 1976	633	12.83	
Wellesley, Mass.	Post, 1962	3,014	13.2	
Chicago, Illinois	Abernethy, 1925	487	13.5 ± 1.1	16.5–17 +
Moosehart, Illinois	Reymert, 1940	138	13.1	
Louisiana	Gould & Gould, 1932	1,037	13.74 ± 0.03	9–19
Minnesota, Minneapolis	Treloar, 1974	324	13.6	10–19
New York, New York	Popenoe, 1928	200	13.6 ± 0.06	
New York, New York	Lintz & Markow, 1923	800	13.5	11–18
New York, New York (Hebrew Orphan Asylum)	Boas, 1932	185	13.5 ± 1.1	
New York, New York (Horace Mann)	Boas, 1932	116	13.1 ± 1.2	
New York, New York (HM non-Hebrew)	Boas, 1932	236	13.1 ± 1.2	
New York, New York	Engle & Shelesnyak, 1934	250	13.5	
New York, New York Whites	Michelson, 1944	1,922	13.08	
New York, New York Blacks	Michelson, 1944	1,397	13.40	
New York, New York Whites (middle class)	Hillman et al., 1970	2,992	12.9	
Cleveland, Ohio	Simmons & Greulich, 1943	200	12.6 ± 1.1	11.1–15.7
Philadelphia, Pennsylvania	Mazer & Israel, 1951	3,014	13.2	
Yellow Springs, Ohio	Reynolds & Wines, 1948	32	12.9 ± 1.4	10–16
NE, SW, NW, S Whites	McMahon, 1973	10,291	12.80	6–17

Table 23 (Continued)

Population	Investigators	Number	Age at Menarche	Range
NE, SW, NW, S Blacks	McMahon, 1973	8,685	12.52	6–17
South (poor)	Burrell et al., 1961	27,036	15.42 ± 0.04	
South	Burrell et al., 1961	20,834	15.02 ± 0.03	
South Richmond, Virginia Blacks	Mills & Ogle, 1936	108	13.6 ± 0.13	
All parts of U.S. (students, nurses)	Zacharias & Wurtman, 1968	6,217	12.65	
Midwest & Rio de Janeiro	Eveleth, 1966	262	12.60 ± 2.0	
American, Australian & English, Singapore	Wadsworth, 1953	27	13.0	
Obese	Wolff, 1955	47	13.6	
Alaska, Eskimo, Point Barrow	Engelmann, 1903		13.0	
Alaska, Eskimo	Levine, 1953		14.4	
Assam	Rakshit, 1960	138	12.4 ± 0.99	9–15
Upper, Digboi (Privileged)	Foll, 1961	1,150	13.2 ± 1.6	
Kachari	Deka, 1976	200	12.3 ± 0.28	
Burma, Upper, Chauk (Privileged)	Foll, 1958	702	13.2 ± 1.1	
Upper	Foll, 1961	1,150	13.2 ± 1.6	
Canada, Montreal	Jenicek & Demirjian, 1974	1,002	13.08 ± 0.04	
Ceylon, Rural	Wilson & Sutherland, 1953	296	14.39 ± 1.73	
Colombo City	Wilson & Sutherland, 1953	844	12.84 ± 1.24	
Colombo City	David & Jaywardene, 1974	661	13.72 ± 0.12	
Chile, Santiago	Rona & Pereira, 1974	354	12.6	
China, Canton	Chau & Wright, 1925	2,291	14.5	
Hong Kong	Mills & Ogle, 1936	284	15.82 ± 0.06	
Hong Kong (rich)	Lee, Chang, & Chan, 1963	278	12.5	
Hong Kong (average)	Lee, Chang, & Chan, 1963		12.8	
Hong Kong (poor)	Lee, Chang, & Chan, 1963		13.3	

Chinese, Eurasian, Indian Vietnamese Singapore	Wadsworth, 1953	58	13.6	
Cuba, Havana Blacks	Pospisilova et al., 1965		12.9	
Havana Whites	Pospisilova et al., 1965		13.0	
Havana Mulattoes	Pospisilova et al., 1965		13.0	
Denmark, Copenhagen	Bojlén et al., 1954	528	13.6 ± 1.13	
England, Birmingham	Waddy, 1846	623	14.2	
Birmingham	Hogben et al., 1948	662	13. 6 ± 0.07	
Southern	Wilson & Sutherland, 1950	2,590	13.49 ± 1.91	9–18
Sussex	Wilson & Sutherland, 1950	202	13.11 ± 0.09	9–17
Yorkshire	Wilson & Sutherland, 1950	87	13.82 ± 0.14	
Sussex (not rich)	Wilson & Sutherland, 1960	1,029	13.03 ± 0.06	
Sussex (rich)	Wilson & Sutherland, 1960	283	13.30 ± 0.18	
Dorset (not rich)	Wilson & Sutherland, 1960	1,151	13.38 ± 0.05	
Various parts of the country	Ellis, 1950	470	13.73 ± 1.37	
Various parts of the country Oxford (blood donors)	Grant et al., 1956	516	13.32	
Various parts of the country London (middleclass)	Scott, 1961		13.06 ± 0.02	
Various parts of the country Durham (all classes)	Roberts et al., 1971	1,608	13.43 ± 0.05	
Finland	Malmio, 1919	1,219	15.56	
	Simell, 1952	3,075	14.3 ± 1.1	
	Salo, 1947	3,630	13.76	
Formosa	Ko, 1933	3,000+	13.2	
France, Paris	Bourliére et al., 1966	277	12.10	
Germany, Berlin	Grimm, 1952	3,000+	13.2	
	Zeller, 1944	230	12.9	
Bavaria (upper class)	Stratz, 1922		12.9	
Bavaria (middleclass)	Stratz, 1922		14.4	
Bavaria (peasants)	Stratz, 1922		16.4	

Table 23 (Continued)

Population	Investigators	Number	Age at Menarche	Range
Holland	Van't Land & de Haas, 1957	1,500	13.6 ± 1.0	
	Bolk, 1923	1,800	13.9	11–19
	Rusback et al., 1961	1,800	13.7 ± 0.06	
	Van Veurooy, 1976	577	13.37 ± 0.09	
Hungary, Budapest	Thoma, 1960	414	12.75 ± 0.08	11–16
	Farkas, 1962		13.00	
	Bottayan et al., 1963	7,008	13.2 ± 0.02	10–16
Iceland, Reykjavik	Magnüsson, 1978	682	13.06 ± 0.10	8.5–17+
India, Bombay	Robertson, 1845	230	13.6	
Bengal, Brahman	Sen, 1953	153	12.80 ± 0.10	9–16
Bengal, Vaidyas	Sen, 1953	275	12.70 ± 0.08	9–16
Bengal, Kayasthas	Sen, 1953	219	12.87 ± 0.08	10–16
Bengal, Nayars	Sen, 1953	60	14.29 ± 0.19	10–16
Bengal, Bagdis	Sen, 1953	74	13.25 ± 0.15	11–16
Central, Munda, Oraon, Santal	Wilson & Sutherland, 1953	411	14.65 ± 1.55	
	Israel, 1959	2,227	13.42	
Bengal	Banerjee & Mukherjee, 1961	1,047	13.60	
West Bengal	Mukherji & Sengupta, 1962	230	13.6	
India, South, Madras (urban)	Madhavan, 1965	248	12.76	11–18
South, Madras (rural)	Madhavan, 1965	219	14.16	12–18
Kerala (urban)	Madhavan, 1965	253	13.24	12–18
South, Kerala (rural)	Madhavan, 1965	226	14.42	12–18
South, Warangal & Madras (poorest)	Roberts, et al., 1977	1,267	14.08	
South, Warangal & Madras (poor)	Roberts et al., 1977		13.74	
South, Warangal & Madras (most advantaged)	Roberts et al., 1977		12.86	

Iraq, Baghdad	Shakir, 1971	4,663	13.6	
Israel	Sohar & Rieber, 1960	1,000	13.6	
Tel Aviv	Ber & Brociner, 1964		13.00	
Italy, Florence	Young et al., 1963	111	12.5	
Tuscany	Prosperi et al., 1965	2,256	14.6	
Naples	Carfagna et al., 1972	8,261	12.49 ± 0.02	
Veneto	Gallo, 1977	6,252	12.77 ± 0.03	
Jamaica, Blacks (poor)	Tom & Tom, 1966	1,000	14.85	10–21
(nurses)	Tom & Tom, 1966	391	13.55	10–18
Japan	Makuda & Hori, 1939	375	14.8	11–20
	Yamasaki, 1909	4,861	14.8	9–21
Tokyo 1953 (students)	Matsumoto et al., 1963	616	14.9	12–18
Tokyo 1961 (students)	Matsumoto et al., 1963	806	12.9	12–18
Gumma 1959 (students)	Matsumoto et al., 1963	1,850	13.10	15–18
Gumma 1959 (farmers)	Matsumoto et al., 1963	213	15.03	14–50
	Kimura & Kitano, 1959		14.03	
Lapps	Kiil, 1939		16.5	
Madeira	Robertson, 1846	228	15.5	11–20
Malaysia, Singapore	Aw & Tye, 1970	3,328	12.69	
Martinique, Creole	Benoist, 1963	350	14.06	13–19
Mexico, Mérida, Urban	Diaz Bolio, 1967	993	12.61 ± 0.08	
Villa de Xochimilco, D. F. Urban	Diaz de Martman et al., 1968	670	12.76 ± 0.07	
Mexico City, D.F. Urban	Peña Gomez, 1970	400	12.75 ± 0.10	
Tampico, Urban	Peña Gomez, 1970	416	12.55 ± 0.10	
Tampico-Altamiro, Rural	Peña Gomez, 1970	212	13.79 ± 0.20	
Oaxaca, Rural and Urban	Malina et al., 1977	315	14.27 ± 0.20	
New Guinea, Bundi-Chimbu	Malcolm, 1966		17.6 ± 0.28	
Bundi non-Chimbu	Malcolm, 1966		18.6 ± 0.33	
Bundi (all clans)	Malcolm, 1970		18.0 ± 0.19	
New Guinea, Chimbu	Malcolm, 1966		17.5 ± 0.36	
Kaiapit	Malcolm, 1969		15.6 ± 0.25	

Table 23 (Continued)

Population	Investigators	Number	Age at Menarche	Range
Megiar	Malcolm, 1969		15.5 ± 0.20	
Lumi	Wark & Malcolm, 1969		18.4 ± 0.36	
Norway	Skerlj, 1939	3,408	14.56 ± 0.03	
	Jacobson, 1954	288	14.2 ± 0.1	
Oslo	Kiil, 1953		13.4	
Panama, Blacks	Mills & Ogle, 1936	128	14.04 ± 0.11	
Philippines	Mills & Ogle, 1936	106	14.6 ± 0.14	
Poland, Wroclaw	Zukowski, 1964	1,621	12.6 ± 0.05	11–16
Wroclaw	Milicer, 1968	5,859	13.21 ± 0.02	9.8–17.7
Warsaw	Milicer & Szczotka, 1966	3,129	13.01 ± 0.00	11–19
Rural	Laska-Mierzejewska, 1970	9,589	14.04 ± 0.02	
Scotland, Edinburgh	Kennedy, 1933	10,219	15.03 ± 1.70	
Slovenia, NE	Kralj-Čerček, 1956	223	13.61 ± 0.88	10–17.5
N. Adriatic	Kralj-Čerček, 1956	52	13.94 ± 0.17	
Sweden	Lennér, 1941	2,000	14.48 ± 0.03	
Stockholm	Furu, 1976	917	13.09 ± 0.08	
All	Romanus, 1952	2,348	13.71	
	Bengt-Olov et al., 1974	360	12.94 ± 1.06	
Turkey, Istanbul	Olcay et al., 1975	1,468	12.36 ± 0.01	
Wales, Swansea	Roberts & Dann, 1975		12.79	
Yucatan	Steggerda, 1941	42	12.91	
Yugoslavia, Ljubljana	Skerlj, 1950	1,709	13.41 ± 0.02	

See References, end of chapter.

Table 24
Average Age at Attainment of Menarche in Different Groups[50]

Country or People	No Groups Assembled	Total No Cases	Range Group Average Age at Menarche (years)
Cochin China	1	1,244	16.6
Denmark	3	7,391	15.6–16.8
Sweden	2	5,548	15.2–16.6
Hottentot-Bastards	1	19	15.7
Germany	10	46,221	14.8–16.4
Negroes, Barbados	1	77	15.6
Austria	4	5,695	14.8–16.2
Russia	9	37,854	14.4–16.5
Hungary	1	9,600	15.4
Czechoslovakia	2	10,003+	15.0–15.5
England	6	21,683	14.96–15.6
Norway	8	6,072+	14.3–16.1
Finland	6	44,420	14.4–16.0
Madeira	2	468	14.8–15.5
Poland	2	18,623	15.1–15.2
Armenia	2	5,000	15.0–15.2
Korea	1	424	15.1
Ainu	2	183	15.0–15.2
Japan	19	38,146	14.3–15.8
Spain	1	8,000	14.9
France	6	3,489	14.1–15.5
Netherlands	3	2,084	13.8–15.3
Italy	3	33,493+	13.7–15.0
China	5	8,875	13.5–15.0
U. S. A. Whites	6	26,163	13.7–14.8
U. S. A. Negroes	2	2,339	13.2–14.1
American Indians	5	1,168	12.0–15.0
India	4	1,569	12.5–13.6
Totals	117	345,851+	12.04–16.80

ADAPTED from G. Frommolt, *Rassenfragen in der Geburtshilfe und Gynäkologie*. Leipzig, J. A. Barth, 1936.

groups are given in Table 23. The range of group averages of age at menarche in 117 groups from 28 different regions of the earth is given in Table 24.

An interesting study of adolescent sterility in India was made in 1950 by Miss Tulika Sen under the direction of Dr. S. S. Sarkar of the Indian Museum at Calcutta. Altogether

318 women are reported upon. The mean age of the upper caste Bengali girls (Brahmans, Vaidyas, Kayasthas, collected from two Calcutta girls' schools) was 12.78 ± 0.05 years (s.d. 1.27, v. 1.61). Sen examined the data on 120 women of these three castes who were mostly married before menarche. The adolescent sterility interval in these 120 women appears to have been 29.30 ± 1.96 months or two years and five months. In 53 rural Bagdi women, who married before menarche, the adolescent sterility interval was 31.21 ± 2.07 months or two years and seven months.[51] In 12 urban upper caste women who married within six months following menarche the adolescent sterility interval was 37.83 ± 5.92 or three years and two months. These data are set out in Table 25.

Mandelbaum, in a valuable paper on the fertility of women in the early years of marriage in India, shows that effective intercourse begins appreciably later than is generally supposed, and that the average age of the mother at birth of her first child is rarely less than 17 years and more likely to be nearer 19 years.[42] Mandelbaum concludes that "the presently available evidence indicates an average menarche-first birth interval of the order of three years when effective marriage begins within a few months of menarche."

Data presented by J. N. Sinha relating to fertility of marriage by age at marriage in Lucknow and Kanpur indicate longer intervals between effective marriage and first conception among women married before 19 years than at later ages. This is shown quite clearly in Table 26.[43]

In 1965 P. P. Talawar reported a study of 2713 Bengalee women from three districts of Calcutta. Marriage was consummated at ages ranging from nine to 22 years. The findings, which speak for themselves, are set out in Table 27

Since the mean age of these women at menarche was nearly 14 years the mean adolescent sterility interval was some two years.[44]

Geoffrey Gorer has interestingly recorded the conditions of reproductive life among the Himalayan Lepchas, a

Table 25
Menarche-Conception Intervals

Subjects	Interval in months	Frequency
	Below 10	16
Upper caste	11–20	41
Bengali women	21–30	25
	31–40	8
	41–50	8
	51–60	6
	61–70	8
	71–80	4
	81–90	3
	91–100	1
	Total	120
	Mean 29.30 ± 1.96 months Variance 476.00 S. D. 21.40	
Rural Bagdi women	2	4
	14	8
	26	16
	38	12
	50	12
	62	1
	Total	53
	Mean 31.21 ± 2.07 months Variance 225.56 S. D. 15.02	
	9	1
Upper caste	13	1
women	15	2
	17	1
	19	1
	29	1
	39	1
	56	2
	58	1
	68	1
	Total	12
	Mean 37.83 ± 5.92 months Variance 421.52 S. D. 20.53	

SOURCE: Sen, *The Reproductive Phases of Some Indian Women*, 1950, Tables 5, 7, 8.

Table 26
Age-First Live Birth Intervals

Age at Gauna*	Number	Mean Interval Gauna-1st Live Birth (years)
Under 13, or "prior to puberty"	128	5.0
13–15	273	3.2
16–18	305	2.5
19–21	166	1.7
22–24	57	1.7

*"Gauna" refers to consummation of marriage, assumed to establish actual conjugal relations.
SOURCE: J. N. Sinha, "Fertility and Age at Marriage." Document P 43/Nov 16, International Statistical Institute, New Delhi, Session, 1951.

Table 27
Adolescent Sterility Among Calcutta Women

No	Age at Consummation	Mean Sterility Interval	Median Age 1st Birth
186	9	6.36	15.36
220	10	5.53	15.53
354	11	4.29	15.29
404	12	3.47	15.47
351	13	2.71	15.71
356	14	1.96	15.96
250	15	1.67	16.67
196	16	1.44	17.44
113	17	1.33	18.33
105	18	1.12	19.12
70	19	0.92	19.92
51	20	1.03	21.03
25	21	1.35	22.35
32	22	0.89	22.89

Mongoloid people living in Sikkim. This primitive group forms a scattered minority throughout much of Sikkim and in the very north of Bengal. Only in Zongu, a 220-square-mile reserve containing about 2,000 inhabitants in the Northwest Sikkim, is there an exclusively Lepcha community. Gorer found that "though copulation normally starts at or before puberty the average age for bearing the first child (worked out from the women of Lingthem) is over twenty-two; there are only three instances of a woman bearing a child before she was twenty. Most exceptionally for a primitive community it was possible in Zongu to get the exact age of every individual" owing to the care that they must take in this matter in order to get their horoscopes properly cast.[45]

In a later chapter Gorer writes, "Marriages take place . . . when the parties are still very young, and there is normally a considerable lapse of time between marriage and the birth of the first child. Only two women in Lingthem [the first village in which Gorer worked] bore children before they were eighteen and the greater number usually had their first child somewhere around the age of twenty."[45]

In an appendix Gorer gives the year of birth of the mother and the age of the mother in round years at the birth of her first child. From these figures the average age of the mother at birth of first child is found to be 22.4 years.

The Lepchas neither recognize nor have a word for physical puberty, most of them being of the belief that the signs of puberty in the female are the result of sexual intercourse, and that it is therefore necessary for girls to have intercourse if they are to develop. Girls are betrothed and occasionally married from the age of eight onwards, and boys from the age of twelve, and from this time intercourse may take place regularly between them. There is, however, no age requirement for the commencement of sexual activities, and children indulge in sexual play from an early age. Nevertheless, in spite of the fact that contraceptives are completely unknown, and there exists no cultural necessity for the practice of abortion, since neither "illegitimate" children nor large families are regarded unfavorably, pregnancy rarely occurs before the twenty-second year.

Boys, according to Gorer, generally produce their first emissions from about the age of fifteen, and menstruation appears to occur in girls at about an average age of thirteen years. Thus, the duration of the adolescent sterility interval in the Lepcha female appears to be of the order of nine years, the longest interval yet recorded for any group of humans. This extraordinarily protracted interval is very probably in some manner connected with the high rate of female total sterility that occurs among the Lepchas. Thus, out of a total of 59 women, shown in Table 28, 23 individuals or 39 percent were sterile. Gorer specifically makes the point that of the sterile women under sixty he is certain that they had never borne a child either alive or dead. Thirty-nine percent is an extremely high sterility rate, and an average of nine years is an abnormally long period of adolescent sterility to occur in a group of women in whom the age at birth of first child was known, out of a total of 36 in twelve of whom age at birth of first child was unknown. It seems highly probable that the adolescent and total sterility rates are, in the case of the Lepchas, closely connected, an association that is in itself of considerable interest.

Nicora, in a paper published in 1938, reports that out of 390 pregnant Italian girls ranging in age from 11 to 17 years, only one girl became pregnant before the commencement of menstruation, while two others had menstruated for only two and six months respectively prior to conception; the remainder did not become pregnant until a more appreciable period of time had elapsed following menarche.[46] It is not stated, however, when active intercourse commenced in each case.

In one of the finest ethnological studies of a social institution of a nonliterate people to be published in this century, namely, of the *ghotul* or village dormitory of the Muria people of Bastar, in the Central Provinces of India, Verrier Elwin has given an admirable account of the facts relating to the infertility of adolescent girls.

Table 28
Fecundity of Lepcha Women

Date of Birth	Live Children M	Live Children F	Dead Children No Conception M	Dead Children No Conception F	Age at Birth of First Child
1852	1	1	1	?	?
1858	1	?	?	?	?
1863	2	4	1	1	?
1864	?	?	?	?	?
1866	1	?	?	?	?
1870	1	—	1	—	?30
1871	—	—	—	—	—
1871	—	—	—	—	—
1873	—	1	2	—	?
1875	2	1	1	—	?
1875	—	—	—	—	—
1877	—	—	1	1	?
1877	—	—	—	—	—
1881	—	3	1	—	?
1883	2	3	1	—	19
1884	—	—	2	—	?
1885	2	3	—	—	?
1887	3	1	—	—	24
1888	3	2	—	—	20
1889	—	—	2	2	?
1890	—	—	—	—	—
1891	—	—	—	—	—
1891	—	—	—	—	—
1891	—	—	—	2	?20
1892	—	—	—	—	—
1893	—	3	3	3	22
1894	—	—	—	—	—
1896	1	—	sb	—	23
1896	—	—	—	sb	—
1896	2	—	?2	?3	?25
1897	—	—	—	—	—
1899	1	4	—	—	22
1899	2	1	1	—	20
1900	—	—	—	—	—
1900	1	2	sb	—	23P
1901	—	2	1	1	?25
1902	—	—	—	—	—
1905	2	1	—	—	19
1906	1	—	?	?	?29
1906	1	3	—	—	17
1906	—	—	—	—	—
1908	1	2	—	sb	22
1909	—	—	—	—	—
1909	—	—	—	—	—

Table 28 (Continued)

Date of Birth	Live Children M	Live Children F	Dead Children No Conception M	Dead Children No Conception F	Age at Birth of First Child
1909	—	3	—	—	22
1909	—	2	1	—	21
1909	—	—	—	—	—
1911	—	1	—	—	25
1911	—	—	—	—	—
1911	—	—	—	—	26
1912	1	—	—	—	22
1912	—	—	—	—	—
1912	1	—	—	—	20
1912	1	—	—	—	22
1914	1	—	sb	—	20P
1914	—	—	—	—	—
1918	—	—	—	—	—
1919	—	—	—	—	—
1920	—	—	—	—	—

sb = stillborn child. P = pregnant in 1937.
SOURCE: Gorer, Appendix I, Table III.

It is the custom among the Muria for boys and girls, from about the age of six years, to leave their parental hearths and make a home of the *ghotul*. Here they take unto themselves a partner of the opposite sex who then becomes their ghotul mate, or in another type of ghotul there is a rotation of mates. Here sexual play commences, and intercourse occurs as soon as it becomes possible. The members of the ghotul generally live there for a good number of years following puberty.

Of 2,000 men examined, 720 had grown up in the *jodidār* ghotul, that is the ghotul where one had a permanent mate, and 1,280 grew up in the ghotul where partners had to be changed. These men copulated at least twice a week. Of the 80 who impregnated girls during their ghotul period, only 25 were living in the *jodidār* ghotul, and 55 in the others, yielding a figure of one pregnancy in 23.2 for the modern ghotul and 1 in 28.8 for the *jodidār*.[47] In other words, probably somewhat less than 4 percent of the adolescent girls exposed to these men in the ghotul conceived.

Table 29
Number of Births for Each Period of Months or Years after Marriage in 2,000 Muria Females

Time After Marriage	No of Births	Time After Marriage	No of Births
Within a mo	1	1 yr	187
After 2 mo	1	1.6 yr	50
3 mo	6	2 yr	463
4 mo	3	2.5 yr	93
5 mo	1	3 yr	603
6 mo	17	4 yr	180
7 mo	10	5 yr	116
8 mo	4	Uncertain	112
9 mo	3		

SOURCE: Elwin, *The Muria and Their Ghotul*, 1947.

The gods that watch over the ghotul save the girls from conceiving "for three years after the menarche," Elwin comments, "There must be centuries of observation behind this faith in a sterility interval."[47] Indeed, the Muria have recognized the existence of a sterility interval but have attributed it to supernatural causes rather than to natural ones. In any event, in recognizing the existence of a sterility interval the Muria were ahead of most civilized peoples. The reason for this is almost certainly to be found in the institution of the ghotul, which allowed for a more clear-cut expression and recognition of the infecundity of its members, as contrasted with the adult population outside the ghotul.

It is best to give Elwin's findings in his own words.

"I give below a table [Table 29] showing the number of births of a first child at varying ages. It may be said with a good deal of confidence that effective marriage* never or rarely begins before a year after the menarche, and in many cases not for two or three years. It never occurs in the prepubertal period.

"Table 29, then, gives the number of births for each period of months or years after effective marriage. The total

*By "effective marriage" Elwin means the permanent marriage generally contracted with a spouse outside the ghotul, in other words, legal marriage.

Table 30
Ages of Girls at First Premarital Pregnancy in 327 Cases

Age	Frequency	Age	Frequency
15	1	20	46
16	26	21	25
17	49	22	18
18	68	23	23
19	63	24	8

SOURCE: Elwin, *The Muria and Their Ghotul*, 1947.

number of cases examined was 2,000, of which 150 were infertile.

"I have some confidence in the accuracy of these figures because the length of time before a child is born is a matter of the greatest interest to both wife and husband. It is probably one of the few things in Muria life about which accurate calculations are made.

"I have less confidence in my other tables, but I give them because they are interesting and suggestive. They represent the result of an inquiry in which men belonging to 220 different ghotul were consulted about the conception problem during their own chelik [unmarried youth] membership in the ghotul periods. Each man may be assumed to have spent at least five years in the ghotul. Few of them showed any reluctance to talk about the affairs of their youth: the statements of those who did were checked by

Table 31
Years Between Menarche and Pregnancy in 327 Cases Cited in Table 28

Years	Frequency	Years	Frequency
1	2	5	30
2	61	6	28
3	99	7	3
4	82	Uncertain	22

SOURCE: Elwin, *The Muria and Their Ghotul*, 1947.

separate inquiries from other members of the same ghotul. In the 220 ghotul there were histories of 327 pregnancies. The Muria were consulted about the ages of the girls who became pregnant, and the number of years which had passed in each case between menarche and pregnancy. Muria calculations are obviously rough and ready, but it is hard to see how we can get more accurate figures for a primitive society. I give them for what they are worth."

Elwin goes on to say:

"Even though these figures cannot be regarded with the confidence one would give to laboratory material, they are strikingly suggestive of the sterility interval. Table 29, which refers to the postmarital period, is confirmed by the two other tables which describe premarital conceptions.

"I have little doubt that here we have the real solution to our problem."[47]

Dr. Henry Greist, who was superintendent of the Presbyterian Hospital at Point Barrow, Alaska, from 1920 to 1936, informs me that among the Eskimo of that vicinity although puberty and intercourse commenced at a relatively early age (between 10 and 12 years)* pregnancy before 15 years of age was extremely rare. Contraceptive practices were entirely unknown.

It is of great interest to note that as early as 1855 John Simpson, surgeon in the Royal Navy, observed of the Point Barrow Eskimos, "The age at which women are married is probably in general fifteen or sixteen. They do not commonly bear children before twenty; and there is usually an interval of four years or more between births."[48]

The interval between births was probably due to the fact that Eskimo mothers generally suckle their young quite intensively for three or four years.

Richards and Reining have reported that among the Bahaya of northern Tanzania girls marry between 15 and 18 years. Menarche occurs at about 13.6 years, and the average age at first pregnancy is 19.1 years.[49]

---

*These ages are almost certainly underestimates.

PRECOCIOUS CONCEPTION

It is well known that many females are capable of bearing children at a chronological age of less than 12½ years, and that they may menstruate long before they have reached that age. In many cases the chronologic age at menarche is unusually low, but there can be little doubt that in normal cases the *physiologic* age, that is to say, the actual developmental status of such girls as measured against chronologic time, is commensurate with that of the chronologically older girls.[50-54] Arbitrary chronologic standards of age, it must be clear, are only of secondary importance in cases of girls who, at eight or nine years of age, have borne children are comparatively few, though in gross number numerous enough. While such cases are the exceptions in terms of chronologic age, there is little doubt that they are in no way exceptional in terms of physiologic age. A girl of less than 12½ years of age may be as sexually mature as a woman of 25, and a woman of 25 may be as sexually immature as a normal girl of less than 12½ years of age. Such cases are obviously abnormal, but the state of development encountered in them does serve to explain why girls of the former class are able to conceive at so early a chronologic age.

We must be careful to avoid the error of expressing qualitative aspects of development in terms of quantitative units of measurement. Chronologic time is really a standardization device that is reified and taken as if it really governed these rates of development, whereas the contrary is true. Standards of expected physiologic (as well as social) behavior are very roughly derived from levels of development and then given chronologic status.

Let us consider a simple example of this practice. A mother, looking through the chronologic frame, may say, "Mary is ten years old today." We correctly understand that, by the calendar, Mary was brought into the world just ten years ago. In the afternoon Mary's physician calls (he happens to be a competent biologist), and makes his quarterly examination. He notes in his record, "Mary Brown, physiologic age eighteen years, chronologic age ten years." Here the physician is observing through an auxiliary frame in addition to the chronologic one. In terms of development Mary, from the physician's standpoint, is eighteen years of age, although chronologically she is simply ten years of age. From the developmental point of view Mary has accomplished in ten years what, on the average, it takes eighteen years to accomplish. By every biological and psychological test Mary is eighteen years of age, the equal of any eighteen-year-old girl who has reached the eighteen-year-old standard of physiological and mental development. But chronologically and legally she is still ten years of age. Should Mary happen to give birth to a child shortly after her tenth birthday, the fact is likely to be widely reported in the press as a sort of natural wonder. From such a chronologic standpoint the immediately relevant fact that from the standpoint of reproductive capacity Mary happens to be of nubile age is of no importance whatsoever. Needless to say, from the point of view of an intelligent society the latter fact is of far greater importance than the former.[55]

The case of Lina Medina, which was reported throughout the world press in May, 1939, is a case in point. It is of particular interest because it is well authenticated, and because Lina is the youngest mother of whom we have a record.

Lina Medina is reported to have been born 23 September 1933, in the little village of Pauranga in the Peruvian highlands. She commenced to menstruate at the age of 8 months, and at 4 years of age developed pubic hair. At the reported age of five years and eight months, on 14 May 1939, she gave birth, at term, to a male infant. She was delivered by caesarian section by Dr. Gerardo Lozoda at the Maternity Clinic of Lima (Peru). The child weighed nearly six pounds (2,700 grams) and was perfectly well formed and developed. The mother's predelivery weight was 67 pounds. Mother and infant left the clinic in excellent health after a few days.[56]

A biopsy of one of the ovaries of Lina showed that the

ovarian development of this chronologically five-year-old girl was that of a physiologically mature woman.[57] Chronologically Lina may have been 4 years and 10 months of age when she conceived, but physiologically she was evidently considerably older, at least four times her chronological age.[58]

It has been erroneously stated that Lina Medina's sexual precocity was due to developmental defects such as are associated with Albright's syndrome. The fact is that Lina Medina never at any time exhibited the slightest evidences of such a condition.[57]

At this late date there appears to be some doubt as to the exact age of Lina Medina at the birth of her child. The world press reported the age as five years and eight months. In a report of another precocious childbirth at the Lima Maternity Hospital *Time* magazine (16 December 1957) gave that as her age at *conception* and the age at which she became a mother as six years five months. I wrote *Time* pointing out that if the birth date of Lina Medina were accurate the age at which she became a mother as given by *Time* could not be correct. Furthermore, since *Time*'s reported age at motherhood disagreed with that given by the world press, I asked *Time* to provide me with its source of information. *Time* at once checked with its sources in Peru and on 14 February 1958 wrote me that "Dr. Gerardo Lozoda, who delivered Lina and who, as you know, has been the protector of Lina and her boy, assures us that she was five years, eight months when she became pregnant and six years, five months when she became a mother."*

My own direct inquiries to Dr. Lozoda and Dr. Escomel met with no response. The reader may, therefore, take his choice as to the age at which Lina Medina became a mother.

In 1941 Brunner gave an account of the outcome of 1,556 conceptions in 979 women living in or in the vicinity of

---

*Letter to the author from Ms. Marie Barry for the editors of *Time,* 14 February 1958.

New York; he found that the smallest number of conceptions and the greatest number of spontaneous abortions occurred in females below the age of 20. Brunner observes that "a relative infertility of the adolescents may account for both the small number of conceptions and the large percentage of spontaneous abortions."[59] About 23 percent of spontaneous abortions occurred in females less than 20 years of age, a percentage exceeded only in the females above 36 years of age. Exactly what the conception rate was we are not told, except that it was lower than in any other age group.

Between the years 1936 and 1940 Professor Morris Steggerda made a careful study, among other things, of the sexual history of the females in two villages in Yucatan, his object being to check on such groups the present writer's theories with respect to adolescent sterility. In 1941 Professor Steggerda published the report of his findings. The villages in which the work was carried out are known as Piste and Pencuyut. As among most rural communities of Yucatan, an unusual custom prevails in these two villages. The girls discontinue school at the time of their first menstruation. By this means Steggerda was able to determine the mean age at menstruation of 42 girls for whom the exact birth dates were available. This was found to be 12.91 years.

At puberty these Yucatecan girls are eligible for marriage, but this is usually deferred for a few years, the average age at marriage being 15.7 years. Steggerda found that the average age of 33 Piste mothers at the birth of their first child was 17.6 years. "On the average Piste mothers produce their first baby 1.9 years after the average age of marriage. . . . For some of the persons . . . we were able to find marriage dates, which indicated that they had their first babies less than nine months after marriage. These and others, known to have had promiscuous relations prior to marriage, were omitted from consideration and the remaining suitable cases are listed in Table 52 [here Table 32]. . . . Thus we can say that for about eight months following marriage conception does not occur; the Piste mothers marry

Table 32
Interval Between Marriage and Birth of First Child for Piste and Pencuyut Women

Name	Marriage Date	Date 1st Child	Interval Between Marriage-First Child
		PISTE	
F. D.	10/23/20	9/28/21	11 months
C. C.	2/25/34	7/ 1/35	16 months
F. T.	6/13/31	10/23/32	16 months
M. D. M.	6/ 5/28	11/ 4/29	17 months
M. M.	2/13/32	12/ /33	22 months
Y. M.	6/15/32	6/ 6/34	24 months
A. T.	5/12/25	6/ 1/27	25 months
J. E.	4/29/27	11/ 9/29	30 months
V. C.	1/ 9/18	10/14/21	45 months
Average			22.88 months
		PENCUYUT	
D. M.	7/29/19	4/30/20	9 months
P. T.	6/18/32	3/13/33	9 months
A. Y.	4/14/30	2/ 8/31	10 months
M. P. C.	12/ 3/28	10/13/29	10 months
M. M. C.	9/23/33	8/13/34	11 months
E. A.	11/24/34	11/25/35	12 months
M. A. O.	8/ 1/28	8/ 4/29	12 months
M. A. C.	3/20/24	4/20/25	13 months
D. C.	6/12/15	7/11/16	13 months
M. J. M.	4/ 6/25	5/18/26	13 months
G. C.	5/26/24	9/ 3/25	15 months
Y. Y.	2/11/24	7/ 1/25	17 months
F. O.	2/13/24	8/18/25	18 months
M. I. P.	10/ 9/31	5/29/33	20 months
Y. C.	7/18/21	5/24/23	22 months
J. Z.	1/21/33	9/24/35	34 months
Average			14.75 months

ADAPTED from Steggerda, Table 52, p. 213.

earlier, have their first babies at an early age, and have a longer sterility interval than the Pencuyut women. These data agree with Ashley Montagu's theory that there is generally a period of sterility following menarche.''[60]

An interesting investigation designed to test the theory of adolescent sterility is Beebe's study published in 1942.

The study was carried out in Logan County, West Virginia, in a region of low economic opportunity and high population pressure. Since many of the subjects examined were married at ages 13, 14, and 15 years there was an excellent opportunity here to test the theory of adolescent sterility. Beebe's findings are set out in Table 33. As will be seen from the final column of that table, and as Beebe remarks, "the values [for median months between marriage and first conception] fall from 7.5 for women married at thirteen to 2.5 for women married at twenty-two or older."[61] From the horizontal line of totals in Table 33 it will be seen that a similar trend is exhibited for the median number of months between menarche and the first conception, the values declining with increase in age from 4.5 months at 11 years or under to 2.9 months at 17 years or over. The trend is not, however, so clearly marked in the individual age groups.

In this connection Dr. Beebe writes, "At least a partial confirmation of the theory [of adolescent sterility] would ensue if among women married at age 15, for example, one

Table 33
Median* No Months Between Marriage and 1st Conception for Women who Refrained from Contraceptive Practice, by Age at Marriage and Age at Menarche, Logan and Rural Kentucky

	Age at menarche							
Age at Marriage	11 or under	12	13	14	15	16	17 or over	Total
13		6.5	8.5					*7.5*
14	5.8	5.5	*6.0*	9.3	8.5			*6.4*
15	5.3	*7.5*	*3.6*	5.3	3.5			*4.6*
16	3.5	*3.4*	*3.3*	*5.3*	*3.5*	1.2		*3.6*
17	4.5	*2.3*	*4.0*	*3.2*	*4.5*	4.0		*3.2*
18	[1.2]	*3.8*	*2.6*	*3.4*	*2.9*	5.0	6.5	*3.1*
19	5.5	1.5	*2.0*	*2.7*	3.0	5.0	4.0	*2.7*
20–21	5.0	3.0	*3.5*	*1.5*	*2.4*	*3.0*	2.5	*2.5*
22 and over			*2.0*	4.5	2.0	2.3	2.8	*2.5*
Total	*4.5*	*3.3*	*3.2*	*3.6*	*3.0*	*3.8*	*2.9*	*3.34*

* Each median is based on 8 or more cases, and those that are italicized upon 20 or more, except the median 1.2 shown in brackets, which is based on 6 cases.

found that women who first menstruated at this age required a long time for conception, women who were 14 at menarche somewhat less, and that only women as young as 13 at menarche conceived promptly. The medians of Table 33 not only fail to decline appreciably as one moves from left to right across any horizontal line, but they reveal no reliable trend within each age-at-marriage group when studied in relation to age at menarche.''[61]

Dr. Beebe's reasoning here is based on a misunderstanding of the theory of adolescent sterility. Dr. Beebe appears to have been under the impression that the theory of adolescent sterility postulated ''an average delay of one to three years,'' between menarche and the ability to conceive (personal communication, 28 October 1942). While this would appear to be the case in some human groups, and that in others this average is even exceeded, it is clear from an examination of the evidence that variability among different human groups in this respect is considerable, so that any such limiting generalization embracing the human species as a whole would be bound to be incorrect. What the theory of adolescent sterility postulates, and what the evidence fully confirms, is that while a certain proportion of females are able to conceive immediately following the establishment of menarche and even before, a still larger proportion are incapable of doing so until some months, and in some cases years, have elapsed after menarche. Furthermore, there is no reason to believe that females who attain menarche at 12½ years are more likely to mature earlier than those arriving at menarche at 15 years. Such females may mature at the same age and arrive, incidentally, at the menopause at the same age (unpublished investigation by the author).

The fact is that Beebe's findings fully support the theory of adolescent sterility advanced in this book, and Dr. Beebe agrees that his data are not in conflict with that theory (personal communication, 28 October 1942). What his particular findings demonstrate is that there is a considerable amount of variability in the duration of the menarche-conception interval as between different individuals as well

Table 34
Mean Interval (years) Between Age at Marriage of Mexican Mothers and Their First-born Child

Age Range	U.S.A. Mothers	U.S.A. Daughters	Mexico Mothers	Mexico Daughters	Total Mean	Range
12–14	(12) 2.25	(2) 2.00	(7) 1.00	— —	1.95	1–7
15–19	(111) 1.75	(35) 1.94	(64) 1.34	(35) 1.63	1.65	1–12
20–24	(32) 1.72	(25) 1.60	(44) 1.52	(24) 1.62	1.61	1–6
25 +	(9) 1.34	(4) 1.25	(8) 1.38	(10) 1.90	1.52	1–4

ADAPTED from Goldstein, Table 11, p. 27.

as in relation to the age at which menarche occurs. These are matters that deserve further investigation.

Another investigation calculated to throw some light on the problem of adolescent sterility in man is that carried out by Goldstein on Mexican workers and their Texas-reared children. Goldstein writes, "the relationship between age at marriage and first birth is considered in Table 11 [Table 34 here]. Taking the series as a whole, with better numerical representation in each group, it would seem that the earlier the marriage, the greater the interval before the birth of the first child: an average of 1.95 years when married at 12 to 14 years; of 1.52 years when married at 25 years or after. This tendency supports the findings of Ashley Montagu pertaining to the general occurrence of a greater or less period of adolescent sterility after menarche."[62]

Yet another study designed to throw some light on the reality of adolescent sterility is that conducted by Professor Meyer Fortes at Agogo, Ashanti, Gold Coast (now Ghana). Some 1,017 women were the subject of the study. Professor Fortes writes:

"The tremendous value attached to childbearing and the absence of birth control means that a first pregnancy is eagerly looked forward to by both spouses and their kinsfolk. Indeed, if it is too long delayed the marriage may be in danger of breaking up. It is of interest, therefore, to know the average length of the interval between commencement of cohabitation in marriage and first pregnancy.

As premarital sexual freedom is, however, not uncommon among Ashanti girls of childbearing age, and as pregnancy sometimes results, the question should rather be put in the form: what is the mean interval between the onset of menstruation and first pregnancy? Further interest also attaches to this question in view of the contention that there is a period of adolescent sterility in females between the onset of puberty and the establishment of reproductive capacity. The results of our inquiry are given in Table 42 [Table 35 here], but as has already been mentioned, an error of three to six months is highly probable. According to Table 42 [Table 35 here] the mode is for the first pregnancy to start in the second year after the beginning of menstruation. The mean interval is 1.9 years (S.D., 1.7 years) for the whole group, including those who had not yet had a pregnancy. But the variation is considerable, the maximum interval admitted being nearly eight years. These figures must be taken with great caution, but it does not seem unwarranted to infer that an interval of roughly two years between the onset of menstruation and the achievement of first pregnancy is typical; and this would imply perhaps a year of 'adolescent sterility.' Inspection of the table, moreover, suggests that the modal interval is constant for all age groups. This not only strengthens the inference that there exists a period of adolescent sterility, but also suggests that this is related to physiological conditions during the past 40 years. It is apparent, also, why young married women and their husbands become worried if there is no pregnancy within two or three years after marriage.

"Ashanti themselves say that menstruation commences in the sixteenth year, that marriage usually follows within the following twelve months, and that a woman's first child will be born to her within the next two years. Our data suggest that these judgments are correct. It seems safe to say that the normal age of menarche among the Ashanti is about 15.5 years, that first marriage takes place at about 16.5 to 17 years, and that the first pregnancy follows at about 17.6 to 18."[63]

Table 35

Interval Between Menstruation and First Pregnancy in West African (Mostly Ashanti) Girls

Interval (years)	Age Distribution of Women, in Five-Year Groups														
	15–19	20–24	25–29	30–34	35–39	40–44	45–49	50–54	55–59	60–64	65–69	70–74	75–79	80	Total
No pregnancy	64	26	27	9	4	2	5	1	2	1	0	1	0	1	143
Under 1	10	19	17	4	19	4	6	9	6	7	3	3	1	0	108
1–2	26	73	56	48	33	14	26	23	18	16	8	9	3	7	360
2–3	12	32	38	19	19	18	12	10	9	7	5	9	5	5	200
3–4	3	7	18	5	18	8	10	4	4	5	7	2	1	2	94
4–5	0	7	9	2	3	1	3	3	1	2	1	1	0	1	34
5–6	0	4	6	2	0	3	3	4	1	2	0	1	0	0	26
6–7	0	2	2	1	0	2	1	0	1	2	1	0	0	0	12
7–8	0	0	9	2	4	2	2	2	0	3	0	1	0	0	25
Not stated	0	0	4	1	2	1	2	2	1	0	0	0	1	1	15
Total	115	170	186	93	102	55	70	58	43	45	25	27	11	17	1,017

SOURCE: Fortes, *A Demographic Field Study in Ashanti*, 1955.

Table 36
Mean Age at Menarche, First Conception, and Duration of Menarche-Conception Interval in Various Groups

Groups, Region, Investigator, Date	No Cases	Mean Age Menarche	Mean Age Conception	Mean Menarche-Conception Interval
Annamite, Cochin-China (Mondière) 1880	980	16.4	19.8	3.4
Cambodian, Cochin-China (Mondière) 1880	96	16.10	21.8	4.10
Chinese, Cochin-China (Mondière) 1880	106	16.6	18.0	1.6
Chinese, South Fukien (Maxwell) 1909	99	16.1	19.2	3.1
Chinese, Canton (Chau & Wright) 1925	596	14.5	19.7	5.2
Chinese, Hong Kong (Mills) 1934	284	15.8	20.2	4.6
Filipinos, married (Mills) 1934	84	14.	20.10	6.5
Filipinos, unmarried (Mills) 1934	22	14.7	20.10	6.3
Fuegians, Tierra del Fuego (Bove) 1883	—	12.2	16.2	4.0
Hindus, Calcutta (Gupta) 1848	25	12.3	14.1	1.10
Hindus, Bengal (Sen) 1950	216	12.78	16.25	3.47
Hindus, Bombay and Madras (Balfour) 1927	6580	14.21	18.3	4.1
Lepchas, Sikkim (Gorer) 1938	—	13.0	21.7	8.7
Micronesians, Romonum (Gladwin) 1950	—	14.0	19.0	5.0
Min-Huong, Cochin China (Mondière) 1880	62	16.9	19.11	3.2
Muria, India (Elwin) 1947	305	15.79	19.35	3.56
Ashanti (Fortes) 1954	1017	15.5	17.6	2.1
unmarried, Cincinnati (Mills) 1924	33	13.4	17.2	3.9
unmarried, Richmond, Va. (Mills) 1934	56	13.4	17.4	4.0
married, Richmond, Va. (Mills) 1934	52	13.7	20.8	7.1
unmarried, Panama (Mills) 1934	72	14.0	18.4	4.4

Table 36 (Continued)

Groups, Region, Investigator, Date	No Cases	Mean Age Menarche	Mean Age Conception	Mean Menarche-Conception Interval
married, Panama (Mills) 1934	56	14.0	21.0	7.0
Melanesians, Schouten Islands (Hogbin) 1935	50	17.0	20.2	3.0
Panamanians, unmarried, Panama (Mills) 1942	132	14.0	19.2	5.2
Panamanians, married, Panama (Mills) 1942	18	14.2	19.4	5.2
Papuans, Sepik District (Kaberry) 1941	—	13.6	16.6	3.0
Yucatecan, Piste, Yucatan (Steggerda) 1941	33	12.9	16.8	3.11
Total, 27 groups. Mean of total means:	10,774	14.4	18.8	4.5

MARRIAGE-CONCEPTION INTERVAL

Comparison of the figures in Tables 32 through 36 will show that the findings of the investigators, each on different population groups, exhibit fairly considerable differences in the duration of the marriage-conception intervals for each age at marriage in the different groups. It is in the present state of our knowledge impossible to say with any certainty to what these differences may be due, but there is pretty good reason to believe that it is unlikely that genetic factors are here significantly operative. It is more probable that environmental factors are accountable for these differences. What the nature of these environmental differences may be it is again extremely difficult to say. Nutritional factors are well known to bear a significant relation to ovulation, fecundity, and fertility.[64-65] And it is now well established that socioeconomic factors play a highly important role in influencing the onset of menarche.[66] Similar factors may be operative in producing the differences in the marriage-conception intervals in the various geographic groups referred to in these pages.

REFERENCES

1. Schultz, A. H., and Snyder, F. F. Observations on reproduction in the chimpanzee. *Bull Johns Hopkins Hospital.* 57: 198, 1935.

2. Suśruta, quoted from M. Gupta, On menstruation among Hindu females. Edited by A. Webb. In *Pathologica Indica, or the Anatomy of Indian Diseases,* 2nd edition, Calcutta, London, and Paris: Thacker, pp. 276–81.

3. Plato. *The Republic,* Book V, Chapter 46. 340 B.C. (?).

4. Aristotle. *Politics,* Book V, Chapter 14. 330 B.C. (?) W. D. Ross, editor. In *The Works of Aristotle.* London: Oxford University Press, 1921, p. 1334.

5. Duncan, J. M. *Fecundity, Fertility, Sterility, and Allied Topics.* Edinburgh: A. and C. Black, 1886, p. 24.

6. Renauldin. Nubilité. *Dictionnaire des Sciences Médicales.* 36: 481–3, 1819.

7. Maygrier, J. P. Menstruation. *Dictionnaire des Sciences Médicales.* 32: 375–96, 1819.

8. Waddy, J. M. Birmingham Lying-In Hospital. Medical and Statistical Report, *Lancet.* 1: 674, 1846.

9. Dubois, P. A. and Pajot, C. *Traité Complet de l'Art des Accouchements,* Vol. 1. Paris: Bechet, 1860, p. 269.

10. Joulin, J. D. *Traité Complet de'Accouchements.* Paris: Savy, 1867, pp. 103–4.

11. Duncan, J. Matthews. On the age of nubility. *Edinburgh Med J.* 12: 207–14, 1866/67.

12. Pillai, N. K. *Census of India, 1931.* Travancore. Vol. 28, part 1. Trivandrum, 1931, p. 151.

13. Pearl, R. *The Natural History of Population.* New York: Oxford University Press, 1939, p. 43.

14. Harris, R. P. Early pregnancy. *Am J Obstet.* 6: 571–5, 1973/74.

15. Pearl, R. *The Natural History of Population.* New York: Oxford University Press, 1939, p. 43.

16. Wilhelmi, C. Manners and customs of the Australian natives, in particular of the Port Lincoln district. *Trans R Soc Victoria.* 5: 164, 1861.

17. Bove, G. Über die Fueurlander. *Globus.* 43: 10, 158, 1883.

18. Grutzner, H. *Berliner Missionsbericht.* Berlin: 1901, p. 114.

19. Himes, N. E. *Medical History of Contraception.* Baltimore: The Williams & Wilkins Co., 1936, pp. 3–56.

20. Devereux, G. *A Study of Abortion in Primitive Societies.* Rev. ed. New York: International Universities Press, 1976.

21. Webb, Allan. *Pathologica Indica, or the Anatomy of Indian Diseases.* 2nd edition. Calcutta, London, and Paris: Thacker, 1848, p. 261.

22. Mondière, A. -T. Sur la monographie de la femme de la Cochinchine. *Bull de la Societe d'Anthropologie,* Series 3, 3: 250–61, 1880.

23. Mondière, A. -T. Nubilite. *Dictionnaire des Sciences Anthropologiques.* 823–4, 1886/94.

24. Hartman, C. G. On the relative sterility of the adolescent organism. *Science.* 74: 226–7, 1931.

25. Mills, C. A., and Ogle, C. Physiologic sterility of adolescence. *Hum Biol.* 8: 614, 1936.

26. Chau, K. T., and Wright, J. M. Gynecological Notes: Canton Hospital. *Chin Med J.* August: 684–7, 1925.

27. Mills, C. A. Climatic effects on growth and development, with particular reference to the effects of tropical residence. *Am Anthropologist.* 44: 1–13, 1942.

28. Mayo, Katherine. *Mother India.* New York: Harcourt, Brace & Co., 1927, p. 22.

29. Robertson, J. On the period of puberty in Hindu women. *Edinburgh Med Surg J.* 64: 156–69, 1845.

30. Robertson, J. Supplement to Mr. Robertson's paper on the period of puberty in Hindu women. *Edinburgh Med Surg J.* 64: 257–64, 1845.

31. Webb, A. *Pathologica Indica, or the Anatomy of Indian Diseases.* Calcutta, London, and Paris: Thacker, 1848, p. 254.

32. Robertson, J. On the period of puberty in Hindu women. *Edinburgh Med Surg J.* 56: 56–64, 1846.

33. Khan, A. H. *Census of India, 1931.* Punjab. Report. 17: 176–80, 1933.

34. Middleton, L., and Jacob, S. M. *Census of India, 1921.* Punjab and Delhi. 15: 244–87, 1923.

35. Clark, A. H. Is India dying? A reply to 'Mother India'. *Atlantic Monthly.* 41: 271–9, 1928.

36. Balfour, M. I. 'Mother India.' Conditions of childbirth. Calcutta. *Times of India.* Vol. 90, October 10, 1927, p. 8.

37. Curjel, D. R. The reproductive life of Indian women. *Indian J Med Res.* 8: 366–71, 1920.

38. Pillay, A. P. Marriage and divorce in India. *Proc 3rd Sexual Reform Cong*. London: Kegan Paul, 1930, pp. 76–90.

39. Sen, T. Reproductive life of some Indian women. *Man in India*. 23: 31–45, 1953.

40. Reche, O. Untersuchungen über Wachstum und Geschlechtsreife bei Melanesichen Kindern. *Korrespondenzblatt der Deutschen Gesellschaft für Anthropologie, Ethnologie, und Urgeschichte*. 41: 48–55, 1910.

41. Ackerley, J. R. *Hindoo Holiday: An Indian Journal*. London: Chatto & Windus, 1932, pp. 96–7.

42. Mandelbaum, D. G. Fertility of early years of marriage in India. *Ghurye Felicitation Volume*. Bombay, 1954, pp. 1–19.

43. Sinha, J. N. Fertility and age at marriage. Document P 43/Nov 16, International Statistical Institute, New Delhi, Session, 1951.

44. Talawar, P. P. Adolescent sterility in an Indian population, *Hum Biol*. 37: 256–61, 1965.

45. Gorer, G. *Himalayan Village*. London: Michael Joseph, 1938, p. 175.

46. Nicora, G. La gravidanza e il parto nelle giovanissime. *Annali di Ostetrica e Ginecologia*. 60: 137–90, 1938.

47. Elwin, V. *The Muria and Their Ghotul*. New York: Oxford University Press, 1947, p. 463.

48. Simpson, J. *Observations on the Western Eskimos and the Country They Inhabit*. London, 1855.

49. Richards, A. I., and Reining, P. Report on Fertility Surveys in Buganda and Buhaya, 1952. Edited by F. Lorimer. In *Culture and Human Fertility*. Paris: Unesco, 1954, pp. 372–3.

50. Dietrich, H. *Die Menarche in ihrer Beziehung zur Menstruation und Fruchtbarkeit des Weibes*. Giessen, 1920.

51. Harris, R. P. Early puberty. *Am J Obstet*. 3: 611–22, 1971.

52. Horman, J. M. J. R. D. *Ueber Menstruatio Praecox*. Leipzig, 1918.

53. Lenz, J. Vorzeitige Menstruation, Geschlechtsreife und Entwicklung. *Archiv für Gynäkologie*. 99: 67–144, 1913.

54. Jolly, H. *Sexual Precocity*. Springfield, Illinois: Charles C Thomas, Co. 1955.

55. Montagu, M. F. Ashley. Social time: a methodological and functional analysis. *Am J Sociol*. 44: 282–4, 1938.

56. Escomel, E. La plus jeune mère du monde. *Presse Médicale*. 47: 875, 31 May 1939.

57. Escomel, E. L'ovaire de Lina Medina. *Presse Médicale*. 47: 1628, 19 December 1939.

58. Dockerty, M. B., et al. Albright's syndrome. *Arch Intern Med.* 75: 372, 1946.

59. Brunner, E. K. The outcome of 1,556 conceptions: a medical and sociological study. *Hum Biol.* 13: 159–76, 1941.

60. Steggerda, M. *Maya Indians of Yucatan.* Washington, D.C.: Carnegie Institution of Washington, Publication No. 531, 1941, p. 212.

61. Beebe, G. B. *Contraception and Fertility in the Southern Appalachians.* Baltimore: Williams & Wilkins, 1942, p. 68.

62. Goldstein, M. S. *Demographic and Bodily Changes in Descendants of Mexican Immigrants.* Austin, Texas: Institute of Latin-American Studies, University of Texas, 1843, p. 27.

63. Fortes, M. A demographic field study in Ashanti. Edited by F. Lorimer. In *Culture and Human Fertility.* Paris: Unesco, 1954, pp. 297–99.

64. Wolfe, B. The fecundity and fertility of early man. *Hum Biol.* 5: 52–4, 1933.

65. Baird, D. The influence of social and economic factors on stillbirths and neonatal deaths, *J Obstet Gyn British Empire.* 52: 217–234, 1945.

66. Mills, C. A. Geographic and time variations in body growth and age at menarche. *Hum Biol.* 9: 43–56, 1937.

REFERENCES TO TABLE 23

1. Fortes, M. A demographic field study in Ashanti. Edited by F. Lorimer. In *Culture and Human Fertility*. Paris: Unesco, 1954, pp 253–324.

2. Burgess, A. P., and Burgess, J. H. L. The growth of some Uganda schoolgirls. *East Afr Med J*. 42: 135–142, 1965.

3. Richards, A. I., and Reining, P. Report on fertility surveys in Buganda and Buhaya. Edited by F. Lorimer. In *Culture and Human Fertility*. Paris: Unesco, 372–373, 1954.

4. Hiernaux, J. La Crossance des Escoliers Ruwandais. Royale Société de Academie des Sciences. Outre-Mer, Brussels. 1965.

5. Hautvast, J. Physical growth and menarcheal age in Tanzanean schoolchildren and adults. *Hum Biol*. 43: 421–444, 1971.

6. Ibid.

7. Oettle, A. G., and Higginson, J. The age at menarche in South African Bantu (Negro) girls. *Hum Biol*. 33: 181–190, 1961.

8. Kark, E. Sexual maturation and variation in the height and weight growth of Bantu girls in Durban. *J Trop Pediatr*. 2: 32–40, 1957.

9. Burrell, R. J. W., Healy, M. J. R., and Tanner, J. M. Age at menarche in South African Bantu schoolgirls living in the Transkei Reserve. *Hum Biol*. 33: 250–262, 1961.

10. Frere, G. Mean age at menopause and menarche in South Africa. *S Afr J Med Sci*. 36: 21–24, 1971.

11. Ibid.

12. Margolius, K. The age of menarche in South African colored girls. *Leech*. 40: 19–22, 1970.

13. Howell, N. The population of the Dobe area ¡Kung. Edited by R. B. Lee, and I. DeVore. In *Kalahari Hunter-Gatherers, Studies of the ¡Kung San and Their Neighbors*. Cambridge: Harvard University Press, 1976, 137–151.

14. Van Castricium, M. Age of menarche on Witwatersrand, South Africa. *South Afr Med J*. 20: 594–595, 1946.

15. Benjamin, F. The age of menarche and the menopause in white South African women and certain factors influencing these times. *South Afr Med J*. 34: 316–320, 1960.

16. Ellis, R. W. B. Age of puberty in the tropics. *Br Med J*. 1: 85–95, 1950.

17. Tanner, J. M., and O'Keefe, B. Age at menarche in Nigerian girls with a rate on their heights and weights from age 12 to 19. *Hum Biol*. 34: 187–196, 1962.

18. Wilson, D. C., and Sutherland, I. The age of the menarche in the tropics. *Br Med J*. 2: 607–608, 1953.

19. Ibid.

20. Oduntun, S. O., Ayeni, O., and Kale, O. The age of menarche in Nigerian girls. *Ann Hum Biol.* 3: 269–274, 1976.

21. Ibid.

22. Gallo, P. G. The age of menarche in Somalia. *Ann Hum Biol.* 2: 197–200, 1975.

23. Kinsey, A. C. et al. *Sexual Behavior in the Human Female*. Philadelphia: W. B. Saunders Co., 1953.

24. Nicholson, Q. B., and Hanley, C. Indices of physiological maturity; derivation and interrelationships. *Child Devel.* 24: 1–38, 1953.

25. Livson, N., and MaNeill, D. The accuracy of recalled age of menarche. *Hum Biol.* 34: 218–221, 1962.

26. Zacharias, L., and Wurtman, R. J. Blindness: its relation to age of menarche. *Science.* 144: 1154–1155, 1964.

27. Damon, A., Damon, S. T., Reed, R. B. et al. Age at menarche of mothers and daughters with a note on accuracy of recall. *Hum Biol.* 41: 161–175, 1969.

28. Shuttleworth, F. K. Sexual maturation and the physical growth of girls age six to nineteen. *Monogr Soc Res Child Devel. 2:* 1–252, 1937.

29. Ibid.

30. Zacharias, L., Rand, W. M., and Wurtman, R. J. A prospective study of sexual development and growth in American girls: the statistics of menarche. *Obstet Gynecol Survey.* 31: 325–337, 1976.

31. Post, R. H. Early menarcheal age of diabetic women. *Diabetes.* 11: 287–290, 1962.

32. Abernethy, E. M. Correlations in physical and mental growth. *J. Educ Psychol.* 16: 458–468, 1925.

33. Reymert, M. L. Relationships between menarcheal age, behavior disorders, and intelligence. *Character & Personality.* 8: 292–300, 1940.

34. Gould, H. N., and Gould, M. R. Age of first menstruation in mothers and daughters. *JAMA.* 98: 1349–1352, 1932.

35. Treolar, A. E. Menarche, menopause, and intervening fecundability. *Hum Biol.* 46: 89–107, 1974.

36. Popenoe, P. Inheritance of age of onset of menstruation. *Eugenical News.* 13: 101, 1928.

37. Lintz, W., and Markow, H. Relation of onset of menstruation to environment. *Endocrinology.* 7: 57–60, 1923.

38. Boas, F. Studies in growth. *Hum Biol.* 4: 307–350, 1932.

39. Ibid.

40. Ibid.

41. Engle, E. T., and Shelesnyak, M. C. First menstruation

and subsequent cycles menstrual of pubertal girls. *Hum Biol.* 6: 431–453, 1934.

42. Michelson, N. Studies in the physical development of Negroes. IV. Onset of puberty. *Am J Phys Anthropol.* 2: 151–166, 1944.

43. Ibid.

44. Hillman, R. W., Slater, P., and Nelson, M. J. Season of birth, parental age, menarcheal age, and body form: some interrelationships in young women. *Hum Biol.* 42: 570–580, 1970.

45. Simmons, K., and Greulich, W. W. Menarcheal age and height, weight, and skeletal age of girls age seven to seventeen years. *J Pediatr.* 22: 518–548, 1943.

46. Mazer, C., and Israel, S. L. *Diagnosis and Treatment of Menstrual Disorders and Sterility.* 3rd Ed. New York: Paul B. Hoeber, 1951.

47. Reynolds, E. L., and Wines, J. V. Individual differences in physical changes associated with adolescence in girls. *Am J Dis Child.* 75: 329–350, 1948.

48. McMahon, B. Age at menarche. United States. *Vital and Health Statistics.* Series 11, No. 133. Washington, D.C.: Department of Health, Education, and Welfare, 1973, pp vi–29.

49. Ibid.

50. Cf Reference 10.

51. Ibid.

52. Mills, C. A., and Ogle, C. Physiologic sterility of adolescence. *Hum Biol.* 8: 607–615, 1936.

53. Zacharias, L., and Wurtman, R. J. Age at menarche. Genetic and environment influences. *N Engl J Med.* 280: 868–875, 1969.

54. Eveleth, P. B. Eruption of permanent dentition and menarche of American children living in the tropics. *Hum Biol.* 38: 60–67, 1966.

55. Wadsworth, G. R. Menstrual characteristics of women living in a tropical climate. *J. Obstet Gynecol Br Empire.* 60: 110–113, 1953.

56. Wolff, O. H. Obesity in childhood: study of birthweight, height, and onset of puberty. *Q J Med.* 24: 109–123, 1955.

57. Engelmann, G. I. The age of the first menstruation at pole and equator. *Am J Gynecol Soc.* 3: 1903.

58. Levine, V. E. Studies in physical anthropology. III. The age of onset of menstruation of the Alaskan Eskimo. *Am J Phys Anthropol.* 11: 252, 1953.

59. Rakshit, S. A short note on the menarcheal age of Assamese girls. *Man in India.* 40: 52–55, 1960.

60. Foll, C. F. The age at menarche in Assam and Burma. *Arch Dis Child.* 36: 302–304, 1961.

61. Deka, R. Age at menarche and haemoglobin E among Kachari women of Upper Assam. *Man in India*. 56: 349–354, 1976.
62. Foll, C. F. Physical development of schoolgirls in Upper Burma. *Arch Dis Child*. 33: 452–456, 1958.
63. Cf Reference 64.
64. Jenicek, M., and Demirijian, A. Age at menarche in French Canadian urban girls. *Ann Hum Biol*. 1: 339–346, 1974.
65. Cf Reference 19.
66. Ibid.
67. David, L., and Jaywardene, C. H. S. Age at menarche in Ceylon. *Ann Hum Biol*. 1: 447–449, 1974.
68. Rona, R., and Pereira, G. Factors that influence age of menarche in girls in Santiago. *Hum Biol*. 46: 33–42, 1974.
69. Chau, K. T., and Wright, J. M. Gynecological notes: Canton Hospital. China Med J. 684–687, August 1925.
70. Mills, C. A., and Ogle, C. Physiologic sterility of adolescence. *Hum Biol*. 8: 608–615, 1936.
71. Lee, M. M. C., and Chang, K. S. F., and Chan, M. M. C. Sexual maturation of Chinese girls in Hong Kong. *Pediatrics*. 23: 389–398, 1963.
72. Ibid.
73. Ibid.
74. Wadsworth, G. R. Menstrual characteristics of women living in a tropical climate. *J Obstet Gynecol Br Empire*. 60: 110–113, 1953.
75. Pospisilova-Zuzaklva, V., Stukovsky, H., Valsik, J. A. The menarche in whites, Negresses, and Mulatto women in Havana. *Z Aertzl Fortbild* (Jena). 59: 500–516, 1965.
76. Ibid.
77. Ibid.
78. Bojlen, K., Rasch, G., and Bentzon, M. The age incidence of menarche in Copenhagen. *Acta Obstet Gynecol Scand*. 33: 405–433, 1954.
79. Waddy, J. M. Birmingham Lying-In Hospital. Medical and Statistical Report. *Lancet*. 1: 674–675, 1846.
80. Hogben, H., Waterhouse, J. A., and Hogben, L. Studies on puberty. Part I. *Br. J. Social Med*. 1: 29–42, 1948.
81. Wilson, D. C., and Sutherland, I. Age at the menarche. *Br Med J*. 1: 1267, 1950.
82. Ibid.
83. Ibid.
84. Wilson, D. C., and Sutherland, I. Present age of menarche in southern England. *J Obstet Gynecol Br Empire*. 67: 320–322, 1960.
85. Ibid.
86. Ibid.

87. Ellis, R. W. B. Age of puberty in the tropics. *Br Med J.* 1: 85–95, 1950.

88. Scott, J. A. *Report on the Heights and Weights of School Pupils in the County of London in 1959.* London: London County Council, 1961.

89. Roberts, D. F., Rozner, L. M., and Swan, A. V. Age at menarche, physique and environment in industrial northeast England. *Acta Pediatr Scand.* 60: 158–164, 1971.

90. Malmio, H. R. *Ueber das Alter der Menarche in Finnland.* Dissertation. Helsingfors, 1919.

91. Simeel, G. A. Uber das Menarchealter in Finnland. *Acta Pediatr.* (Stockholm) 41: (suppl 84), 1952.

92. Salo, A. Suomalaisen kasvatuksen peruskysymyksia. Del. II. Helsingfors, 1947.

93. Ko, K. E. Statistical studies on the menstruation of Formosans. *J Med Assoc Formosa.* 32: 987–1010, 1933.

94. Bourlière, F., Clement, F., and Parot, S. 'Normes' de viellissement morphologique et physiologique d'une population de niveau socio-economique eleve de la region Parisienne. *Bull Mem Soc Anthropol Paris.* 10: 568–577, 1966.

95. Grimm, H. Ein menarchestatistik von Berliner Schulabgangern (1951). *Zentralbl Gynakol.* 74: 1743–1745, 1952.

96. Ibid.

97. Stratz, C. H. *Der Korper des Kindes und seine Phlege.* 9th Ed. Stuttgart: Ferdinand Enke, 1922.

98. Ibid.

99. Ibid.

100. Van't Land, G. M., and Haas, J. H. Menarche-leeftijd in Nederland. *Nederl tijscr v geneesk.* 101: 1425, 1957.

101. Bolk, L. The menarche in Dutch women and its precipitated appearance in the youngest generation. *Koniglike Akademie van Wetenschappen.* Proceedings of the Section on Sciences. 26: 650–663, 1923.

102. Rusbach, H. W., Van Laar, F., and Haas, J. H. Menarche-leeftijd: vergelijking van status quo-en nevraagmethods. *Tidsskr Soc Geneeskunde.* 40: 294–296, 1961.

103. Van Veurooy-Ijsselmiuden, M. E., Meets, M. J. L., and Van der Werff ten Bosch, J. J. The secular trend in age at menarche in the Netherlands. *Ann Hum Biol.* 3: 283–284, 1976.

104. Thoma, A. Age at menarche, acceleration and heritability. *Acta Biol Acad Sci Hung.* 11: 241–254, 1960.

105. Bottayan, O. et al. Age at menarche in Hungarian girls. *Annales Historoico-Naturales Musei Nationalis Hungarici (Pars Anthropologica.)* 55: 562–572, 1963.

106. Magnusson, T. E. Age at menarche in Iceland. *Am J Phys. Anthropol.* 48: 511–514, 1978.

107. Robertson, J. On the period of puberty in Hindu women. *Edinb Med Surg J*. 64: 165–169, 1845.

108. Sen, T. Reproductive life of some Indian women. *Man in India*. 33: 31–54, 1953.

109. Ibid.

110. Ibid.

111. Ibid.

112. Ibid.

113. Cf Reference 19.

114. Israel, S. L. Onset of menstruation in Indian women. *J Obstet Gynecol*. 66: 311, 1959.

115. Banerjee, D., and Mukherjee, S. P. The menarche in Bengalee Hindu girls. *J Indian Med Assoc*. 17: 261–270, 1961.

116. Mukherji, P. S., and Sengupta, S. K. Menarche among school girls in urban west Bengal. *Indian J Child Health*. 11: 119–125, 1962.

117. Madhavan, S. Age at menarche of south Indian girls belonging to states of Madras and Kerala. *Indian J Med Res*. 63:

118. Ibid.

119. Ibid.

120. Ibid.

121. Roberts, D. F., Chinn, S., Girija, B. et al. A study of menarcheal age in India. *Ann Hum Biol*. 4: 171–177, 1977.

122. Ibid.

123. Ibid.

124. Shakir, A. The age of menarche of girls attending schools in Baghdad. *Hum Biol*. 43: 265–270, 1971.

125. Sohar, E., and Rieber, E. Age of menarche in girls of various Jewish communities in Israel. *Harefuah*. 59: 303–305, 1960.

126. Ber, A., and Brociner, C. Age of puberty in Israeli girls. *Fertil Steril*. 15: 640–647, 1964.

127. Young, H. B., Zoli, A., and Gallagher, J. R. Events of puberty in a group of 111 Florentine girls. *Am J Dis Child*. 106: 568–577, 1963.

128. Prosperi, P., Borselli, L., and Matucci-Cerinic, L. Indagini statistica sui più precoce sviluppo sessuale delle ragaze in Toscana. *Riv di Clin Pediat*. 74: 28–45, 1965.

129. Carfagna, M. E., Figurelli, G., and Materese, S. Menarcheal age of schoolgirls in the district of Naples, Italy, in 1969–70. *Hum Biol*. 44: 117–125, 1972.

130. Gallo, P. G. The age of menarche in some populations of the Veneto, north Italy. *Ann Hum Biol*. 4: 179–181, 1977.

131. Tom, F., and Tom, P. The age of menarche in Jamaica. *West Indian Med J*. 15: 83–88, 1966.

132. Ibid.

133. Makuda, T., and Hori, K. Menarche and sterility. *Jpn J Obstet Gynecol.* 22: 190–194, 1939.

134. Yamasaki, M. Ueber den Beginn der Menstruation bei Japanerinnen. *Zeitschr Gynakol.* 33: 1296, 1909.

135. Matsumoto, S., Ozawa, M., Nogami, Y. et al. Menstrual cycle in puberty. *Gumma J. Med Sci.* 12: 119–143.

136. Ibid.

137. Ibid.

138. Ibid.

139. Kimura, K., and Kitano, S. Growth of the Japanese physique in four successive decades before World War II. *J Anthropol Soc Nippon.* 67: 721, 1959.

140. Kiil, V. Stature and growth of Norwegian men during the past 200 years. *Skr Norske Vidensk Akad.* 6: 175, 1939.

141. Robertson, J. On the age of puberty in the island of Madeira. *Edinb Med Surg J.* 65: 281–285, 1846.

142. Aw, E., and Tye, C. Y. Age of menarche of a group of Singapore girls. *Hum Biol.* 42: 329–336, 1970.

143. Benoist, J. Les Martiniquais: anthropologie d'une population métisée. *Bull Mem Soc d'Anthropol Paris.* 4: 341–342, 1963.

144. Diaz Bolio, J. E. Influencia de la nutrición en el desarrollo pubertal de un grupo de adolescentes del sexo feminino de la cuidad de Mérida, Yuc., (*a*) Edad de la menarquia. *Boletin Medico del Hospital Infantil de Mexico,* 21 (Supplemento num. 1), 1967, 119–129.

145. Diaz de Marthman, C. D., Lando Rica, V. M., and Ramos Galvan, R. Crescimento y desarrollo en adolescentes femininos, (2) Edad de la menarquia. *Boletin Medico del Hospital Infantil de Mexico,* 25: 1968, 775–786.

146. Peña Gomez, R. M. *Edad de la menarquia en tres groupos de niñas mexicanas.* Mexico, D. D., Instituto Nacional de Antropologia e Historia, 1970.

147. Ibid.

148. Ibid.

149. Malina, R., Chumlea, C., Stepnick, C. C., and Lopez, F. G. Age of menarche in Oaxaca, Mexico schoolgirls, with comparative data for other areas of Mexico. *Annals of Human Biology,* 4: 1977, 551–558.

150. Malcolm, L. A. Age of puberty in the Bundi people. *Papua New Guinea Med J.* 9: 15–20, 1966.

151. Ibid.

152. Malcolm, L. A. Growth and development of the Bundi child of the New Guinea Highlands. *Hum Biol.* 42: 293–328, 1970.

153. Cf Reference 134.

154. Malcolm, L. A. Growth and development of Kaiapit

children of the Markham Valley, New Guinea. *Am J Phys Anthropol.* 31: 39–51, 1969.

155. Ibid.

156. Wark, L., and Malcolm, L. A. Growth and development of the Lumi child in the Sepik district of New Guinea. *Med J. Austr.* 2: 129–136, 1969.

157. Skerlj, B. Menarche und Umwelt. *Z Menschliche Vererbungsund Konstitutions-lehre.* 23: 299–359, 1939.

158. Jacobsen, L. On the relationship between menarcheal age and adult body structure. *Hum Biol.* 26: 127–132, 1954.

159. Cf Reference 53.

160. Ibid.

161. Zukowski, W., Kmietoxicz-Zukowska, A., and Gruska, S. The age at menarche in Polish girls. *Hum Biol.* 36: 233–234, 1964.

162. Milicer, H. Age at menarche in Wroclaw girls, Poland, in 1966. *Hum Biol.* 40: 249–259, 1968.

163. Milicer, H., and Szczotka, F. Age at menarche in Warsaw girls. *Hum Biol.* 38: 199–203, 1966.

164. Laska-Mierzejewska, T. Effect of ecological and socioeconomic factors in the age at menarche, body height, and weight of rural girls in Poland. *Hum Biol.* 42: 184–292, 1970.

165. Kennedy, W. P. The menarche and menstrual type: notes on 10,000 case records. *J Obstet Gynecol Br Empire.* 40: 7920804, 1933.

166. Kralj-Čerček, L. The influence of food, body build, and social origin on the age of menarche. *Hum Biol.* 28: 393–406, 1956.

167. Ibid.

168. Lenner, A. Das menarchealter. *Acta Obstet Gynecol Scand.* 24: 229, 1941.

169. Furu, M. Menarcheal age in Stockholm girls, 1967. *Ann Hum Biol.* 3: 587–590, 1976.

170. Romanus, T. Menarche in schoolgirls: effects of joint and of separate education. *Acta Genetica.* 3: 50–60, 1952.

171. Bengt-Olov, L., Bergsten-Brucefors, A., and Lindgren, G. The secular trend in physical growth in Sweden. *Ann Hum Biol.* 1: 245–265, 1974.

172. Olcay, N., Alp, U., and Orlon, A. Sexual maturation in Turkish girls. *Ann Hum Biol.* 2: 49–59, 1975.

173. Roberts D. F., and Dann, T. C. A 12-year study of menarcheal age. *Br J Prevent Soc Med.* 29: 31–39, 1975.

174. Steggerda, M. *Maya Indians of Yucatan.* Washington, D.C.: Carnegie Institution of Washington, Publication 531, 1941, 202–221.

175. Skerlj, B. *Development of Secondary School Students of Ljubljana.* Ljubljana: Universa V Ljubljana Institutza Antropologijo, 1950.

6

THE BIOLOGY OF ADOLESCENT STERILITY

The evidence cited in the foregoing pages clearly shows that the immature human female who is less than 20 years of age conceives much less easily, as a rule, and much less frequently, in general, than the mature female up to 30 years of age. It is also clear that a high percentage of these younger women are relatively sterile, and that, in general, prior to the age of 20 ± 2 years the process of childbearing constitutes a hazard to the life of both mother and child which increases in proportion to the youthfulness of the mother and decreases to a minimum at maturity.[1-10] All the evidence has pointed strongly in the same direction, namely, to the conclusion that the adolescent female is biologically immature, developmentally unprepared, for the processes of reproduction; that when these processes are early initiated and exercised before maturity is reached the well-being of the mother and of the child is endangered. In young adolescent females there generally exists a colligation of conditions which, in the largest number of cases, tends to secure such females against undergoing an experience for which biologically they are inadequately prepared. This is putting the matter somewhat teleologically; it would, perhaps, be more correct to say that the adolescent organism being itself in process of development is not generally endowed with the capacity to perform the functions of reproduction until it has reached a certain stage of efficient differentiation. This latter stage depends upon the capacity, among other things, to produce mature ova; and this is a capacity which, as we have seen, there is good reason to believe is not attained in most mammals until some appreciable time after the appearance of the first estrus or menarche.

The ontogenetic developmental history of all mammals consists in a process of differentiation in which organ systems undergo differential development, at rates and in regions that follow one another in a definite serial pattern, at more or less definite moments of time as determined by the system in which the developmental process proceeds. The process begins with the fertilization of the ovum and terminates with the attainment of maturity. In the human female the evidence points to the age of 23 ± 2 years as the age of maturity. The optimal conditions for reproduction prevailing in the organism at that time, that is, at the age of 23 ± 2 years, are maintained for about five years, at the end of which time, that is to say between 28 and 30 years of age, the process of involution commences, and at about 45 years of age expresses itself in the menopause. Puberty represents an early transition period in the total development process, a bridge that leads from childhood to maturity, during which maturation gradually takes place in certain parts of the reproductive system, and leads to the phase of development we recognize as maturity, the period when the organism actually becomes capable of efficient reproduction.

DEVELOPMENTAL PHYSIOLOGY OF REPRODUCTION

In what immediately follows we shall discuss the developmental physiology of the primates with particular reference to the human species insofar as it relates to the functions of reproduction. We must determine to what extent, if any, the physiological evidence supports the idea of the existence of a period of adolescent sterility in the primates.

Development consists of growth or increase in size, and differentiation or increase in complexity. At fertilization the process of development is initiated in the ovum by the spermatozoon; there is immediate growth and gradual increase in complexity. Throughout the process of development growth seems to be one step ahead of differentiation. The spermatozoon contributes the determiners of sex in the form of a specific chromosome-gene structure, which eventually leads to the development of one or another particular set of sex characters, according as the chromosomal structure is of the male or the female determining variety. The development of these characters in mammals is brought about by certain internal secretions, believed to be already present at the so-called "sexually indifferent" stage of development of the organism: the stage during which the sex of the organism cannot ordinarily be determined save by reference to its chromosomal pattern.[11-12]

It appears that from a very early stage in development the general somatic differentiation of the tissues of the organism proceed together, inseparably. All development, in short, is sexual development, the distinction between morphologic and sexual development being an arbitrary one. The process of development consists in the organization of tissues for the fulfillment of certain specific functions: structures have no other meaning than this, and apart from living the chief function of every organism is reproduction, and to this end all development would seem to proceed. Tissues and organs undergo development at rates and in regions that are related to one another in circumscribed serial patterns, so that one process or event cannot occur without exercising a specific influence upon others within the system. The general pattern of development is such, throughout the animal kingdom, that the functional qualities of a tissue or an organ are always preceded by growth and increase in complexity in such tissues and organs. The gradual development of the functions of the somatic system are dependent upon the rates of growth and differentiation of the tissues comprising it; and in the individual development of the organism a particular succession of changes must occur before any one of these functions can be established. So it is with the development of the structures and functions devoted to the reproductive system. These do not appear in the newborn infant in a state ready for the performance of the processes of reproduction, but many years must elapse during which the necessary changes gradually occur that eventually render reproductive activities possible.

The outstanding characteristic of development as a spatio-temporal phenomenon is its gradualness.

At birth the organism is in possession of most of the elements necessary for reproduction. Ova are present in the ovaries, and the pituitary contains gonadotropic hormones.[13] The latter when injected into older animals are capable of inducing ovulation in them. Anterior pituitary hormones, as a rule, are incapable of producing ovulation in the newborn animal. As the animal grows, however, ovarian reactivity to the hormones becomes established, and as sexual maturity is approached the amount of hormone necessary to produce a given reaction progressively decreases. While the ovary is capable of reacting to gonadotropic hormones already present in the anterior lobe of the pituitary at a very short time after birth, under normal physiological conditions it does not do so until long afterwards. It has been suggested that this is due to the fact that though the pituitary contains gonadotropic hormones at this early stage, they are not introduced into the blood stream until a later stage of development has been reached. Since such hormones have been recovered from the urine of two- and three-year-old children, and from the sera of fetuses,[18,19] the explanation must obviously be looked for elsewhere.[14-24]

Evidence of a gradually developing sensitivity of the reproductive system to hormonal stimuli is now well established.[20-25]

Studies, extended over several years, on basal waking temperature in adolescent girls by Cooperman from shortly after they had arrived at menarche, indicated that the estrogenic phase of the normal cycle was associated with low temperature, while the luteal phase was associated with higher temperature. The rise in temperature occurring in the middle of the average 27 to 31 day cycle coincided with ovulation. This thermogenic effect is due to the action of progesterone.[23] In the female who ovulates a characteristic biphasic temperature curve is characteristic. When there is no ovulation the temperature does not rise and the curve is monophasic. Rubenstein found a monophasic temperature curve to be typical of the menstrual cycles of girls in their early postmenarcheal years.[24]

Döring corroborated these findings in a study of 1,471 menstrual cycles in 303 girls. Basal temperature studies showed that in the postmenarcheal female the proportion of anovulatory cycles and cycles with an abnormally short luteal phase is abnormally high. Döring also found that the optimal proportion of normal cycles was only achieved in the 26 to 30 year age group.[25]

Thus the high proportion of anovulatory cycles in girls before the age of 20 strongly points to the existence of a normal relative sterility.

Döring's findings are shown in Table 37. They are based on a total of 3,175 cycles.[26]

In order that a female may conceive and bring the process of pregnancy to a successful termination it is necessary that her ovaries should be able to produce mature ova capable of being fertilized, then implanted in the uterus, and there acted upon in such a manner as to cause the fertilized ovum to undergo development up to the stage when the organism is ready for birth. All these processes require the

Table 37
Anovular Cycles, Cycles with Shortened Hyperthermic Phases and Normal Cycles within the Various Age-Groups

		Anovulatory Cycles		Cycles with Shortened Hyperthermic Phase		Complete Cycles	
Age-group	n	Total	%	Total	%	Total	%
12 to 14	168	101	60	50	30	17	10
15 to 17	316	136	43	130	41	50	16
18 to 20	282	76	27	104	37	102	36
21 to 25	287	37	13	75	26	175	61
26 to 30	418	21	5	50	12	347	83
31 to 35	822	58	7	74	9	690	84
36 to 40	640	19	3	102	16	519	81
41 to 45	275	33	12	49	18	193	70
46 to 50	67	10	15	24	36	33	49

'*n*' is the number of cycles in each age-group. (Döring, 1969)

presence in the maternal organism of a number of gradually developed conditions. In the absence of any of these conditions reproduction cannot occur. These conditions come into existence in a regular order in physiologic and chronologic time, before the establishment of which as a functional complex the female cannot reproduce. If she is not yet able to ovulate she obviously cannot be impregnated, and if she is able to ovulate the absence of the luteinizing hormone (LH) or interstitial cell-stimulating hormone (ICSH) as it is interchangeably known may make it impossible for the fertilized ovum to become attached to the uterine wall. It is just these two conditions, ovulation and formation of the corpus luteum, which do not normally become functionally established until sometime after puberty, if in the conventional sense we here take puberty to mean the period of the first estrus or first menstruation.

Van Herwerden, Allen, Hartman, and Corner were the earliest investigators to show that the early menstruations of the female rhesus monkey are normally unaccompanied by ovulation.[27-31] It is thus impossible for the early menstrual rhesus female to conceive. There is every reason to believe that the same conditions hold true for the human female at a similar stage of development.

It is the hypothalamus that controls the secretions of the pituitary hormones. This control is achieved by neurosecretions from the area of the median eminence of the hypothalamus to the pituitary. The hypothalamus is itself subject to the many influences of the higher centers of the brain.[20-21] For each of the pituitary hormones that acts upon target organs the hypothalamus produces both inhibitory factors and releasing factors. Specific inhibitory factors prevent the secretion of significant quantities of particular pituitary hormones, and specific releasing factors act upon the pituitary to release individual pituitary hormones. The action of the inhibiting and releasing factors is largely under the influence of genetic factors in interaction with environmental conditions such as nutrition, light, altitude, and socioeconomic factors.

There is a temporal sequence in which the hypothalamic inhibitory and releasing factors, and the pituitary hormones begin to act in bringing about pubertal, and eventually other, sexual changes. As Dr. Rose Frisch and colleagues have shown in a series of extended studies, menarche appears to be signalled by the attainment and maintenance of a critical minimum weight in relation to height (Figures 3 and 4).[32-42] In American girls the *average* weight at menarche is about 47 kg (103 lb) and the *average* height is about 158 cm. It seems that it is the relative amount of fat in relation to the lean body mass that is important; a *minimum* of about 17 percent fat/body weight being necessary for menarche, and about 22 percent fat/body weight being essential for the maintenance of regular ovulatory cycles at the completion of growth at ages 16 to 18 years. Individual girls attain these fat percentages or fat/lean ratios at varying heights and weights.

During the adolescent growth spurt, which commences in American girls at about nine and one-half years at an average weight of about 30 kg (66 lb) estimations show that

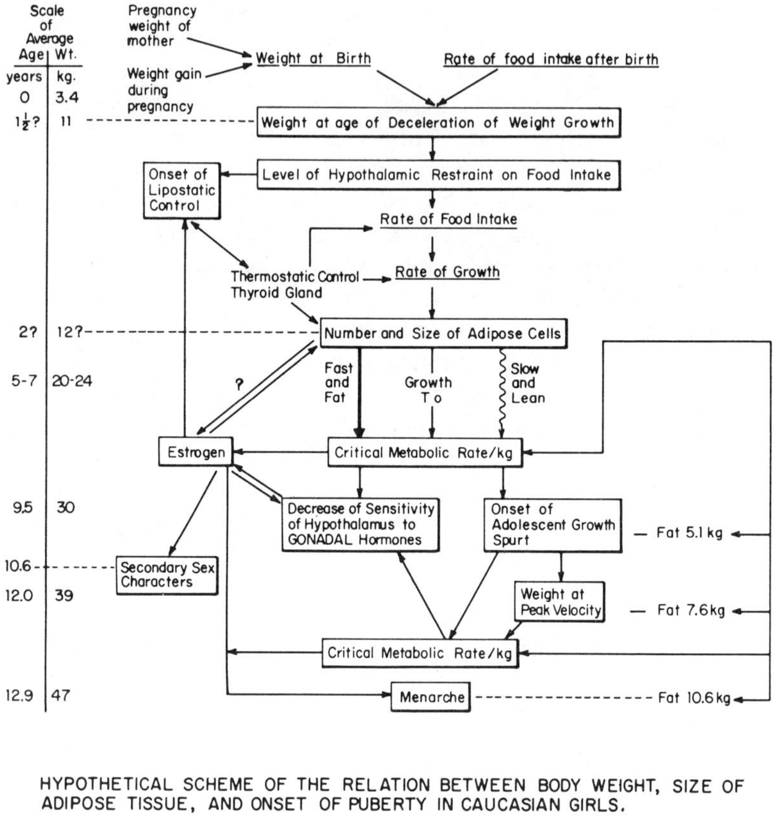

Figure 3. Postulated scheme of the relation between body weight, size of adipose tissue, and the control of the onset of puberty for Caucasian girls. Weight at birth is determined by prepregnancy weight of the mother and weight gain during pregnancy. Food intake per kilogram of body weight after birth determines weight at time of deceleration of weight growth (before or about age 2 years in humans), which sets the level of the hypothalamus restraint on food intake and the accompanying onset of lipostatic control. This, at the same age, determines the rate of food intake, which in turn determines rate of growth and number and size of the adipose cells. Rate of growth and number and size of adipose cells determine metabolic rate per kilogram of body weight. (By permission of Dr. R. E. Frisch and John Wiley & Sons.)

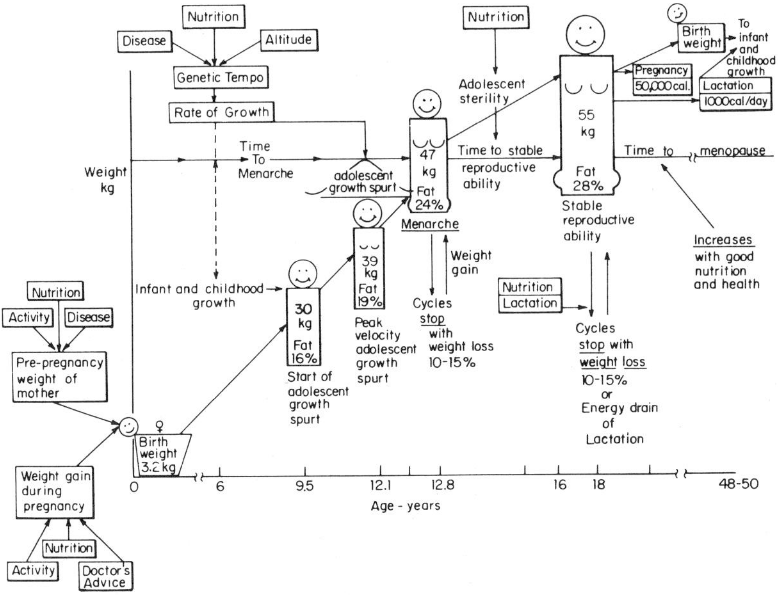

Figure 4. Biological determinants of female reproductive ability (fecundity). Each reproductive milestone can be affected by environmental factors, as shown. The maintenance of regular ovulatory cycles is related to a minimum level of fat storage and is thus directly affected by undernutrition and energy-draining activities, such as lactation. (By permission of Dr. R. E. Frisch and the editor of *Social Biology*.)

stored body fat more than doubles by the time they reach menarche, increasing from growth spurt to menarche by about 120 percent. Body fat at menarche amounts to some 20 to 24 percent of mean body weight. The minimum level of fatness associated with menarche and its maintenance is about 17 percent of body weight. In young girls who are recently menarcheal menstrual function ceases when body weight for a particular height falls below that fatness level. There are, in fact, two minimum weights for a particular height, one at menarche and one after completion of growth at ages 16 to 18 years. The heavier weight for height is associated with the maintenance of regular ovulatory cycles at those ages after growth is completed.

For the maintenance of regular menstrual function a heavier minimum weight for height corresponding to an increased amount of fat/body weight of about 22 percent of body weight is necessary in girls of ages 16 and over. Frisch and McArthur, for example, found that the minimum weight for the resumption and maintenance of regular ovulatory cycles in an 18-year-old girl whose height is 165 cm (55″) is 49 kg (108 lb), whereas the minimum weight for menarche of a girl aged 14 years of the same height is about 43.5 kg (96 lb). The heavier minimum weight for regular ovulatory cycles reflects the increase in fatness that occurs both in early and late maturing girls from menarche to completion of growth at ages 16 to 18 years. The mean fat gain for both early and late maturers from the beginning of the adolescent spurt to menarche is about 6.5 kg (14 lb).[43]

As Frisch points out pregnancy has high energy requirements: 50,000 extra calories are needed to maintain a pregnancy to term, and about 120,000 extra calories are required for lactation over a period of four months, i.e., at the rate of 1,000 extra calories a day. Hence, the period of adolescent sterility is one during which the additional fatness level makes for the attainment of stable reproductive ability and the ability to suckle a child. The level of fatness as represented by body weight of the mother affects the birth weight of the baby.

Frisch and Revelle have suggested that attainment of body weight in the critical range causes change in metabolic rate which, in turn, reduces the sensitivity of the hypothalamus to estrogen, and thus alters the ovarian-hypothalamic feedback.

The evidence from both animal and human studies tends to support this explanation of the manner in which pubertal changes are initiated.[44-63]

What possibly happens at the onset of puberty is this: as the fat level of the body in relation to lean mass reaches a critical phase it activates what is believed to be a gonadotropin stimulating hormone which results in the release of the follicle stimulating hormone (FSH) from the pituitary. The primary function of the follicle stimulating hormone is to instigate morphological development of the entire follicle. FSH causes growth of the ovarian follicle and a filling of its central cavity with follicular fluid containing the hormone estrogen.

When, in experimental animals, the pituitary gland is ablated FSH will not cause the follicles to secrete estrogen. It is probable that FSH is secreted in both sexes continuously with a minor rise in the female in the follicular phase, and a sharp increase just before ovulation.

Estrogen is not only responsible for the gradual appearance of the secondary sexual characters but also for the induction of all those changes in the reproductive tract which result in menstruation. Ovulation, however, does not and cannot occur as a consequence of the action of estrogen alone. At the onset of puberty estrogen is alone present in appreciable amount. For ovulation to occur another factor of the hypothalamus as well as a hormone of the pituitary must come into play: namely, the luteinizing releasing hormone (LRH) and LH.

One current theory has it that the amygdala of the temporal lobe exerts an inhibiting influence on the hypothalamic centers which release LRH, and that it is the gradual decline of this inhibitory influence which is responsible for the onset of puberty. In passing it may be noted that since the amygdala, among other things, monitors emotional stimuli, it has been suggested that stressful situations may produce suppression of LRH, so that luteinizing hormone is not released, resulting in failure of ovulation and amenorrhea.

Another explanation is that the rising concentration of a gonadal steroid like estradiol has a positive feedback on the hypothalamus and causes it to produce LRH. This, in turn, stimulates the pituitary to release large amounts of leutinizing hormones.

The common belief that menarche constitutes the signal of the girl's arrival at the ability to reproduce is in error. Both the development of the secondary sexual characters and the appearance of the first menstruation are evidences of the action of estrogen upon the tissues of the body particularly involved. The girl's menstrual cycles are anovulatory for a period which on the average lasts about three years. During this period the female is quite unable to reproduce. For reproduction to become possible further changes in development must occur resulting in the secretion of LH. Only then does it become possible for ovulation to occur and for a female to conceive.

As we have already seen, this early or primary phase of puberty is of variable duration in any given population of females, and while it may be related to genetic factors[64,65] the expression of these, it is clear, is subject to modification by environmental factors.

Anatomical proof of the anovulatory character of the first menstrual cycles in the human female has been given by Laqueur.[66-68] He described the case of a girl aged 12 years who died from spontaneous intracerebral hemorrhage. This girl was in every way perfectly well developed, and was 5 feet 2 inches in height at death; she had experienced four monthly periods. Examination of the ovaries showed the presence of only one degenerating corpus luteum, thus indicating that there had been three regularly recurrent periods from a healthy uterus without ovulation.

Collett and her coworkers have shown that the highest incidence of anovulatory menstrual cycles in American females occurs during the years from 17 to 19; the length of the cycle is also less constant during these years than at any other time.[69] Again, this evidence indicates that the female in the later years of adolescence is still in process of development as a reproductive organism. At 17 to 19 years of age she has still not attained her mature reproductive development. In Table 38 the incidence of anovulatory cycles in 146 females at various ages is set out.

Ovulation and impregnation become possible only when the second hormone of the pituitary, luteinizing hormone is poured into the blood stream in sufficient quantity and the ovaries reach the proper stage of differentiation or sensitivity. This marks the period of nubility, and as more

Table 38
Incidence of Anovulatory Cycles

Age (year)	Number	Cycles	Anovulatory Cycles Number	Anovulatory Cycles Percent
17–18	59	81	25	30.8
20–24	33	112	11	9.8
25–29	21	45	2	4.4
30–34	11	23	0	0
35–39	13	33	4	12.1
40–50	11	33	5	15.1
All ages	146	327	47	14.3

SOURCE: Collett et al., *The Effect of Age upon the Pattern of the Menstrual Cycle*, 1954.

than one earlier author has pointed out, the period of nubility is not to be confused with the earlier period of puberty. We have already seen, in the nonhuman mammals that have been investigated, that nubility generally follows puberty only after a fairly appreciable amount of time has elapsed. Such evidence as we have for humans indicates that a similar appreciable interval of time must separate the two events or periods in the human species.

Evidence of the gradual development of the gonadotropic excretory activities of the anterior pituitary with the approach to and attainment of adolescence is now available for both sexes. With the refinement of radioimmunoassays both FSH and LH have been demonstrated in the sera of fetuses, infants, and children of both sexes. Considerably higher concentrations of FSH occur in the sera of female than in those of male fetuses. While there is no difference in the concentration of estradiol, there is a highly significant rise in the concentration of testosterone during the second trimester. The level of increase is nine times that of the female. Near term there is a decline in circulating gonadotropins, but this trend is reversed following birth with the rapid decrease in the concentration of the placenta-originating human chorionic gonadotropin (hCG) and estrogens in the serum. As Page et al. suggest, this postnatal surge of gonadotropin in both sexes is probably due to the removal of feedback inhibition by placental steroids.[11]

During the first two years the female shows a rhythmic pattern of FSH release, a rhythm which is repeated in adolescence. Between the ages of three and seven boys and girls continue to excrete a small but constant quantity of estrogens in the urine. From eight to eleven years of age estrogen excretion by girls increases, and about one-and-a-half years before menarche it becomes cyclic, the intensity of the cycles gradually increasing.[14] Such findings indicate that menarche is the culmination of a series of developing neurohumoral conditions that have been proceeding for some years.

Zuckerman had already summarized and stated these facts in 1936.[70] He wrote, "Only the first phase of the full ovulation cycle is experienced for a long period after the process of reproductive maturation (puberty) sets in both in man and monkeys. The anterior lobe of the pituitary presumably undergoes waves of activity during which it elaborates only a follicular-maturation hormone, to cause development of the ovarian follicles. The hormone produced by the developing follicles causes growth of the uterine mucosa, and when the wave of pituitary activity comes to an end, the wave of follicular growth it stimulated also ceases, so that the uterus no longer comes under the influence of the follicular hormone. As a result, a retrogressive uterine phase sets in and menstruation occurs. The first menstrual cycles after puberty probably always follow this course, i.e., the first menstrual cycles experienced by women are usually unaccompanied by ovulation.

"Hence it was that the Trobriand Islanders studied by Malinowski never conceived in their early youth in spite of unrestricted licence. Hence, too, births rarely occur before the third year of effective marital relationship in Indian 'child marriages.' "

Early investigators such as Dorfman, Greulich, and Solomon,[77] Oesting and Webster[78] in investigations on boys and girls, Catchpole and Greulich[79] on girls, and Greulich et al.,[80] on boys had all shown that gonadotropin excretion occurs in small amounts in children three years and over, and that there is a gradual increase in the amount and frequency of occurrence of such hormones with the approach of menarche and maturity.[15,71-73] This has been abundantly confirmed by more recent investigations.

Incidentally, associated with the increase in the intensity of these cycles of estrogen activity before and after menarche are certain other physiological changes. Thus Shock[81] has shown that with respect to pulse rate, blood pressure, and basal metabolism menarche produces no sudden alteration that is reflected in these conditions, but that the rate of change, generally a decrease in rate, occurs in many girls six to 12 months before menarche.[74] "The changes," writes Shock, "in all the physiological functions measured are in the direction of adult values and take place at a more rapid rate after menarche than before." As shown in Shock's curves (Fig. 5), three marked changes are associated with the beginning of menstruation: (1), a leveling off of a premenarcheal rise in systolic blood pressure, (2), a decrease in pulse rate, and (3), a precipitous fall in basal heat production per unit of surface area. Such changes, as Shock suggests, are to be regarded as but another aspect of the process of maturity rather than the result of endocrine changes associated with menstruation.

Puberty, it cannot be too often repeated, is not marked by the first menstruation in the female, nor is it represented by a single characteristic phenotypic change. Puberty is, in fact, a long drawn-out process which begins some two years before the first menstruation, externally announced by the growth of pubic hair and breast development. In the male testis growth and pubic hair appear before the onset of the pubertal spurt in height.[75]

An operationally useful definition of puberty is that offered by Donovan and Van der Werff ten Bosch, as "the phase of bodily development during which the gonads secrete hormones in amounts sufficient to cause accelerated

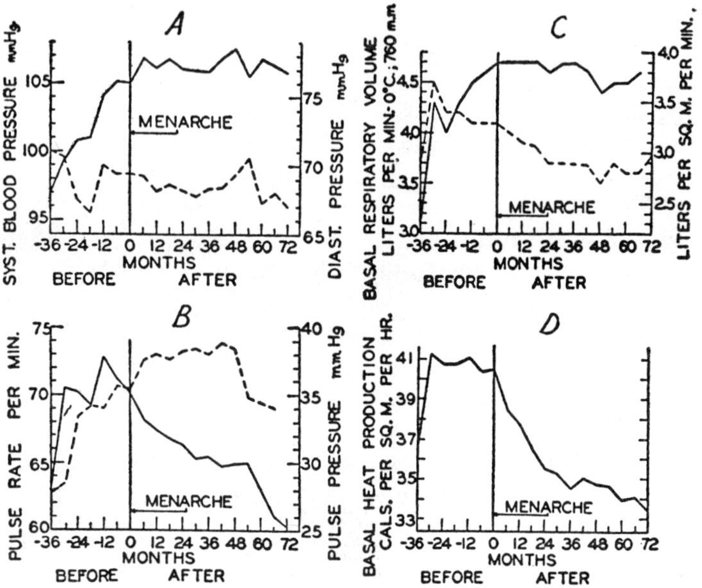

Figure 5. Physiological functions before and after menarche.

A. Effect of menarche on blood pressure in girls. Average values calculated for 6-month intervals before and after menarche for 50 girls. Zero time is taken for each girl at the age when she first menstruated.
——— systolic blood pressure; – – – diastolic blood pressure.

B. Effect of menarche on pulse rate and pulse pressure in girls.
——— pulse rate; – – – pulse pressure.

C. Effect of menarche on basal respiratory volume in girls.
——— liters per minute; – – – liters per square meter per minute.

D. Effect of menarche on basal heat production in girls.
——— basal heat production, calories per square meter per hour.

(After Shock, 1943.)

growth of the genital organs and the appearance of secondary sexual characters."[20] To which I would add, "and the appearance of cyclic bleeding in the female."

As we have seen, much more is involved in the causation of pubertal changes than gonadal secretions, as Van der Werff ten Bosch make abundantly clear. Those changes are, in fact, the effects of the combined operation of genetic, hypothalamic, endocrine, and growth factors. It is not our task to consider those factors in any further detail here; they must however be borne in mind as fundamentally significant in the development of the complex process that puberty represents. The traditional conception of puberty as marked by the first menstruation must be abandoned, for puberty is a complex phase of developmental change which lasts some five or more years.

That an appreciable period of time separates the ability to procreate from the appearance of the first menstruation had already been clearly recognized by Aristotle 2,300 years ago and a century ago by such physicians as Waddy, Joulin, Matthews Duncan, and others, but the scientific world was quite uninterested in the matter. The problem has become of more general scientific interest only in recent years, and it has remained for modern experimental research in the physiology of sexual development and reproduction of other mammals and in humans to rediscover and provide the scientific demonstration of the truths first stated by earlier investigators. In the present chapter an attempt has been made to provide something of the nature of that scientific demonstration.

DISPOSAL OF POSSIBLE OBJECTIONS TO THE THEORY OF ADOLESCENT STERILITY

At this juncture we may examine, and perhaps anticipate, some of the possible objections to the interpretation of the meaning of adolescent sterility that has been given in this book.

It is a well-known fact that many apparently healthy girls have given birth to children at such early ages as eight and nine years, and even earlier, and that many such girls exhibit the characters, at least the physiological characters, of mature women.[76-136] In individuals of the latter variety there is frequently some disorder of the adrenal glands or of the ovaries that is responsible. In many cases where no pathologic process can be demonstrated it is nonetheless possible that such a process is at work, but in many others it seems unnecessary to look for a pathologic factor. What has impressed students of many of these cases of early motherhood and precocious sexual development has been the fact that, as Harris put it in 1873, in these girls their "physical development [is] in correspondence with their sexual" development "and the whole system is in correspondence with it, so that the function [i.e., menstruation] does not, in its performance, interfere with the growth or health of the subject."[137]

Those cases in which sexual precocity is traceable to some abnormal initiation of endocrine activity, which like magic ceases upon the removal of the exciting agent, when they occur in prepuberal or adolescent individuals provide a further illustration of the fact that development toward maturity is dependent upon the functioning of certain organs.[138-141] Such cases help us to understand that in the normal individual the functions of these organs are only gradually established, but that these functions may come into action prematurely as a result of the abnormal excitation of the organs regulating them.

Such cases do not, of course, present anything irreconcilable with the evidence for the existence of an adolescent sterility interval in humans. Finally, it may be added that, while we nearly always hear of the child who has given birth to an infant, we seldom hear of the numerous children who at one time or another have been exposed to sexual attentions from mature males, without ever conceiving. Yet there is no doubt that such cases occur in large numbers in our own and in other societies.

GROWTH AND SEXUAL DEVELOPMENT

An important fact first brought out by Hartman in 1931 in connection with his studies on macaques was that maturity, or the period at which his animals were first able to conceive, was correlated with their general body growth, as particularly reflected in weight. In his animals menarche occurred at an average body weight of 3,000 grams, but ovulation and fecundity were not attained until the average weight of 5,000 grams was reached, thus representing an average sterility interval of about one year.[142] In a Cleveland group of macaques Smith and Rubenstein found the average weight at menarche to be 3,400 grams, while the sterile period varied from one month to a year.[143]

In Formosan girls Quo found that the velocity in growth, as measured by weight, was at its maximum at the onset of puberty, and that following menarche there was a retardation of growth in weight.[144]

Fried and Smith in a study of 408 girls from homes in Cleveland, Ohio, of above average socioeconomic status found that 339 (or 83.1 percent) grew fewer than four inches following menarche. In the majority of females it is apparent from this and other studies that menarche coincides with a deceleration in the rate of growth.[145]

Schultz and Snyder in their study of reproduction in the chimpanzee suggested that the sterility interval is related to the entire growth period in an inverse manner, according to the formula: the shorter the growth period the longer the sterility interval. Thus, these authors wrote in 1935, "Observations available so far, seem to indicate that the period of so-called adolescent sterility is longest in the most primitive form, the macaque, appreciably shorter in the higher primate, the chimpanzee, and shortest, if indeed existing at all, in man. These differences are all the more significant in view of the fact that the entire growth period is shortest in the macaque (about 7 years), much longer in the chimpanzee (11 years) and longest in man (20 years)."[146]

But does it not appear more likely that, if the entire growth period is longest in the higher primates, the periods of infancy, childhood, adolescence, and maturity would be somewhat more extended in duration in them as compared with the lower primates in which the growth period is somewhat shorter? Indeed, this has long been known to be the case, and has been abundantly confirmed by Spence and Yerkes for the chimpanzee as compared with man.[147] It might further be argued that if the adolescent period is of greater duration in the higher primates, that is to say, the period elapsing from menarche to primary fecundity or nubility, then ipso facto the sterility interval should, on the average, be longer in the higher primates than in the lower primates. Certainly it appears clear that the adolescent sterility interval has a far wider range of variability in humans than in the macaque. As for the chimpanzee, the evidence shows that the period varies from four months to more than two years. The chimpanzee evidently resembles humans more closely than it does the macaque, in the present indicated sense, and not in that suggested by Schultz and Snyder. Where the latter appear to have erred is in their original uncritical acceptance of Hartman's statement that "no adolescent female has conceived during the first two years after puberty."[148] We have already seen that the correct figure is actually about one year, as originally given by Hartman in his 1931 communication, while Smith and Rubenstein have demonstrated that the macaque may successfully conceive, and eventually carry a fetus to term, one month after menarche.

Furthermore, the fact that many apparently normally developed females are capable of conceiving immediately after the appearance of menarche, or even before menarche, does not constitute a disproof of the fact that a still larger number of females are incapable of conceiving at these stages of development. If the empirical evidence did not prove it, the facts concerning the physiology of sexual development in the human female would indicate the existence of such an infertile period.

FEMALE CONTROL AND ADOLESCENT STERILITY

Margaret Mead has suggested that the facts of physiology probably constitute only a partial explanation of the phenomenon of adolescent sterility. "Perhaps more important," she writes, "is the shift in the control over the timing of intercourse from the female to the male that comes with marriage."[149] Mead points out that in nonliterate societies before marriage it is the girl who decides whether she will meet her lover, whereas in marriage there is a change. In marriage the husband has access to his wife by right, at any time he chooses. Hence, the inference that Mead draws is that the phenomenon of adolescent sterility is, at least in part, due to the control over intercourse exercised by the unmarried female, since "after marriage, the girls who . . . lived a life of complete freedom conceived and bore children."[149]

The suggestion is an ingenious one, but it happens not to agree with the facts. The error committed by Mead, as the facts from nonliterate societies already cited in this book abundantly prove, is the assumption that girls conceive and bear children after marriage. Of course they do, but not, as a rule, immediately after marriage. The fact is that when girls marry at "puberty" they do not usually conceive for several years after marriage, in spite of the constant sexual attentions of a mature husband. The marriage-conception interval is quite as striking as is the menarche-conception interval, as the studies of Balfour, Elwin, Goldstein, Maxwell, and others cited in this volume show.

Under conditions such as those that exist in the ghotul of the Muria, described by Elwin, where premarital intercourse occurs on an unperiodic regular basis for years, without any control by the female, pregnancies are exceedingly rare. There can be not the least doubt that the physiological facts of adolescent sterility are alone sufficient to account for the observed infertility.

CULTURAL FACTORS

Benedek has discussed the possibility that cultural prohibitions of sexual activities may lead to delay of physiological maturation of the reproductive functions. Benedek has herself pointed out that while inhibiting emotional factors may retard physiological maturation, it is invalid to generalize from this that the stronger the cultural prohibitions, the greater is the delay of sexual maturation. "In this respect," she writes, "the primary constitutional anlage will determine the effectiveness of the emotional factors."[150]

It is, of course, quite possible that in some cases the period of adolescent sterility may be either shortened or lengthened through the influences produced by emotional factors of cultural origin. The disturbing effects of emotional influences upon sexual functions can be very profound.[151-155] The extraordinary duration of the adolescent sterility period, almost nine years, among the Lepchas, for example, may be due to something other than genetic or nutritional factors. But upon this point, insofar as it relates to the phenomenon of adolescent sterility, we have no evidence whatever.

By and large it seems unlikely that cultural or emotional factors have in any way significantly affected the duration of the adolescent sterility interval in most of the populations we have considered.

REFERENCES

1. Duncan, J. M. *Fecundity, Fertility, Sterility, and Allied Topics*. 2nd ed. Edinburgh; A. and C. Black, 1871.

2. Harris, R. P. Early pregnancy. *Am J Obstet*. 36: 571–5, 1873/74.

3. Larini, O., and Giannizi, G. Le funzioni di maternita nelli giovani primpiparae. Sorti di bambini nate da esse. *Ann Ostet Ginecol*. 53: 1177–1234, 1931.

4. Lundh, G. On the problem of age and primiparity. *Acta Obstetrica et Gynecologia*. 4 (suppl): 137, 1926.

5. Lenz, J. Vorzeitige Menstruation, Geschlechtsreife und Entwicklung. *Arch Gynakol*. 99: 67–144, 1913.

6. Munder, F. Ueber die Geburtsverhaltnisse im fruhen Lebensalter. *Arch Gynakol*. 55: 1–18, 1894.

7. Nicora, G. La gravidanza e il parto nelle giovanissime. *Ann Ostet Ginecol*. 60: 137–90, 1938.

8. Webb, A. *Pathologica Indica, or the Anatomy of Indian Diseases*. 2nd ed. Calcutta, London, and Paris: Thacker, 1848.

9. Yerushalmy, J. Neonatal mortality by order of births and age of parents. *Am J Hygiene*. 28: 244–70, 1939.

10. Ewart, R. J. The influence of parental age on offspring. *Eugenics Rev.* 3: 201–4, 1911.

11. Page, E. W., Villee, C. A., and Villee, D. B., *Human Reproduction.* 2nd ed. Philadelphia: W. B. Saunders Co., 1976.

12. Gold, J. J. (editor), *Gynecologic Endocrinology.* 2nd ed. New York: Harper & Row, 1975.

13. Smith, P. E., and Dortzbach, C. The first appearance in the anterior pituitary of the developing pig fetus of detectable amounts of the hormones stimulating ovarian maturity and general body growth. *Anat Rec.* 43: 277–97, 1929.

14. Nathanson, I. T., Towne, L. E., and Aub, J. C. Normal secretion of sex hormones in childhood. *Endocrinology.* 28: 851–65, 1941.

15. Oesting, R. B., and Webster, B. The sex hormone excretion in children. *Endocrinology.* 22: 307–14, 1938.

16. Greulich, W. W. et al. *Somatic and Endocrine Studies of Pubertal and Adolescent Boys.* Monographs of the Society for Research in Child Development. 8: 43–62, 1942.

17. Kulin, H. E., and Reiter, E. O., Gonadotropins during childhood and adolescence: a review. *Pediatrics.* 51: 260–71, 1973.

18. Mishell, D. R., Jr., Nakamura, R. M., and Barderia, J. M., Initial detection of human chorionic gonadotropin serum in normal human gestation. *Am J Obstet Gynecol.* 118: 990–1, 1974.

19. Donovan, B. T., and Van der Werff ten Bosch, J. J., *Physiology of Puberty.* London: Edward Arnold, 1965.

20. Grumbach, M. M., Grave, G. D., and Mayer, F. E. (editors). *Control of Onset of Puberty.* New York: John Wiley & Sons, 1974.

21. Kulin, H. E. The physiology of adolescence in man. *Hum Biol.* 46: 133–44, 1974.

22. Richardson, G. S. Hormonal physiology of the ovary. Edited by J. J. Gold. In *Gynecologic Endocrinology.* 2nd edition. New York: Harper & Row, 1975, pp. 55–77.

23. Cooperman, N. R. The relationship of adolescent menstruation to body temperature and sterility. *Am J Obstet Gynecol.* 57: 701–10, 1949.

24. Rubenstein, B. B. Estimation of ovarian activity by the consecutive-day study of basal body temperature and basal metabolic rate. *Endocrinology.* 22: 41–4, 1938.

25. Döring, G. K. Uber die relative Sterilität in den Jahren nach der Menarche. *Geburtshilfe und Frauenheilkunde.* 23: 30–38, 1963.

26. Döring, G. K. The incidence of anovular cycles in women. *J Reprod Fertil.* 6(suppl): 77–81, 1969.

27. Van Herwerden, M. A. *Bydrage tot de kennis van menstrueelen cyclus en puerperium*. Dissertation. Utrecht, 1905.

28. Allen, E. A. The menstrual cycle of the monkey *M. rhesus:* observations on normal animals, the effects of removal of the ovaries, and the effects of injections of ovarian and placental extracts into spayed animals. *Carnegie Institution of Washington, Contributions to Embryology*. 98: 1–44, 1927.

29. Hartman, C. G. The corpus luteum and the menstrual cycle together with the correlation between menstruation and implantation. *Am J Obstet Gynecol*. 19: 511–22, 1930.

30. Corner, G. W. The relation between ovulation and menstruation in the monkey. *JAMA*. 89: 1838–40, 1927.

31. Corner, G. W. The reproductive cycle of the rhesus monkey. Edited by G. A. Baitsell. In *Science In Progress*. New Haven: Yale University Press, 1951, p. 27–58.

32. Frisch, R. E., and Revelle, R. Height and weight at menarche and a hypothesis of critical body weights and adolescent events. *Science*. 169: 397–9, 1970.

33. Frisch, R. E., and Revelle, R. Height and weight at menarche and a hypothesis of menarche. *Arch Dis Child*. 46: 695–701, 1971.

34. Frisch, R. E., and Revelle, R. The height and weight of girls and boys at the time of initiation of the adolescent growth spurt in height and weight and the relationship to menarche. *Hum Biol*. 43: 140–59, 1971.

35. Frisch, R. E., and McArthur, J. W. Menstrual cycles: fatness as a determinant of minimum weight for height necessary for their maintenance or onset. *Science*. 185: 949–51, 1974.

36. Frisch, R. E., Revelle, R., and Cook, S. Components of the critical weight at menarche and at initiation of the adolescent spurt: estimated total water, lean body mass and fat. *Hum Biol*. 45: 469–483, 1973.

37. Frisch, R. E. Critical weight at menarche, initiation of the adolescent growth spurt, and control of puberty. Edited by M. M. Grumbach, G. D. Grave, and F. E. Mayer. In *Control of Onset of Puberty*. New York: John Wiley & Sons, 1974, pp. 403–423.

38. Frisch, R. E. Demographic implications of the biological determinants of female fecundity. *Soc Biol*. 22: 17–22, 1975.

39. Frisch, R. E., and Revelle, R. Variation in body weights and the age of the adolescent growth spurt among Latin American and Asian population, in relation to caloric supplies. *Hum Biol*. 41: 185, 1969.

40. Frisch, R. E. Critical weights, a critical body composition, menarche and the maintenance of menstrual cycles. Edited by E. Watts, F. Johnston, and G. Lasker. In *Biosocial Interrela-*

tions in Population Adaptation. The Hague: Mouton, 1975, pp. 319–52.

41. Frisch, R. E. Fatness of girls from menarche to age 18 years: with a nomogram. *Hum Biol*. 48: 353–359, 1976.

42. Frisch, R. E. Demographic implications of the biological determinants of female fecundity. *Soc Biol*. 22: 17–22, 1975.

43. McArthur, J. W., O'Loughlin, K. M., Beitins, T. et al. Endocrine studies during the refeeding of young women with nutritional amenorrhea and infertility. *Proc Mayo Clin*. 51: 607–16, 1976.

44. Billewicz, W. Z., Fellowes, H. M., and Hytten, C. A. Comments on the critical metabolic mass and the age of menarche. *Ann Hum Biol*. 3: 51–9, 1976.

45. Johnston, F. E., Roche, A. F., Schnell, L. M. et al. Critical weight at menarche. *Am J Dis Child*. 129: 19, 1975.

46. Osler, D. C., and Crawford, J. D. Examination of the hypothesis of a critical weight at menarche in ambulatory and bedridden mentally retarded girls. *Pediatrics*. 51: 675–9, 1973.

47. Welson, Z., and Bielicki, T. The adolescent growth spurt and the critical body weight hypothesis. *Mater I Prace Antropologie*. 86: 27–33, 1973.

48. Frisch, R. E. Critical metabolic mass and age at menarche. *Ann Hum Biol*. 3: 489–92, 1976.

49. Van 'T Hof, M. A., and Roede, M. J. A Monte Carlo test of weight as a critical factor in menarche, compared with bone age and measures of height, width, and sexual development. *Ann Hum Biol*. 4: 581–5, 1977.

50. Widdowson, E. M., McCance, R. A., and Mavor, W. O. The effect of undernutrition and rehabilitation on the development of the reproductive organs: rats. *J Endocrinol*. 29: 119–26, 1964.

51. McCance, R. A. Food, growth, and time. *Lancet*. 2: 671–5, 1962.

52. Kennedy, G. C. Interactions between feeding behavior and hormones during growth. *Ann NY Acad Sci*. 157: 1049–61, 1969.

53. Kennedy, G. C., and Mitra, J. Hypothalamic control of energy balance and the reproductive cycle in the rat. *J Physiol*. 166: 395–406, 1963.

54. Kennedy, G. C., and Mitra, J. Body weight and food intake as initiating factors for puberty in the rat. *J Physiol*. 166: 408–18, 1963.

55. Widdowson, E. M., and McCance, R. A. Some effects of accelerating growth. I. General somatic development. *Proceedings of the Royal Society, Series B. Biological Sciences* 152: 188, 1960.

56. Guzman, M. A. Impaired physical growth and maturation

in malnourished populations. Edited by N. S. Scrimshaw, and J. E. Gordon. In *Malnutrition, Learning, and Behavior*. Boston: MIT Press, 1968, pp 42–54.

57. Barnett, S. A., and Coleman, E. M. Effect of low environmental temperature on the reproductive cycle of female mice. *J Endocrinol*. 19: 232–40, 1959.

58. Monteiro, L. S., and Falconer, D. S. Compensatory growth and sexual maturity in mice. *Animal Production*. 8: 179–92, 1966.

59. Dickerson, J. W. T., Gresham, G. A., and McCance, R. A. The effect of undernutrition and rehabilitation on the development of the reproductive organs: pigs. *J Endocrinol*. 29: 111–18.

60. Crichton, J. A., Aitken, J. N., and Boyne, A. W. The effect of plane of nutrition during rearing on growth, production, reproduction and health of dairy cattle. I. Growth to 24 months. *Animal Production*. 1: 145–62, 1959.

61. Joubert, D. M. Puberty in female farm animals. *Animal Breeding Abstracts*. 31: 295–305, 1963.

62. Frisch, R. E., Hegsted, D. M., and Yoshinaga, K. Body weight and food intake at early estrus of rats on a high-fat diet. *Proc Natl Acad Sci*. 72: 4172–6, 1975.

63. Frisch, R. E., Hegsted, D. M., and Yoshinaga, K. Carcass components at first estrus of rats on high-fat and low-fat diets: bodywater, protein and fat. *Proc Natl Acad Sci*. 72: 379–83, 1977.

64. Gould, H. N., and Gould, M. R. Age of first menstruation in mothers and daughters. *JAMA*. 98: 1349–52, 1932.

65. Petri, E. Untersuchungen zur Erbbedingtheit der Menarche. *Zeitschrift für Morphologie und Anthropologie*. 33: 43–8, 1935.

66. Laqueur, G. L. Anatomical demonstration of the anovulatory menstrual cycle. *California and Western Medicine*. 63: 268–9, 1945.

67. Isserand-Perrier, M. Etude comparative des certains processus croissance chez les jumeaux. *Journal de Genetique Humaine*. 8: 87–102.

68. Johnston, F. E. Control of age at menarche. *Hum Biol*. 46: 159–71, 1974.

69. Collett, M. E., Wertenberger, G. E., and Fiske, V. M. The effect of age upon the pattern of the menstrual cycle. *Fertil. Steril*. 5: 437–8, 1954.

70. Zuckerman, S. The physiology of fertility in man and monkey. *Eugen Rev*. 28: 39, 1936.

71. Dorfman, R. I., Greulich, W. W., and Solomon, C. I. The excretion of androgenic and estrogenic substances in the urine of children. *Endocrinology*. 21: 741–3, 1937.

72. Catchpole, H. R., and Greulich, W. W. Excretion of gonadotrophic hormone by prepuberal and adolescent girls. *Proc Am Physiol Soc*. 129: 331, 1940.

73. Greulich, W. W., Dorfman, R. I., Catchpole, H. R. et al. Somatic and endocrine studies of pubertal and adolescent boys. *Monographs of the Society for Research in Child Development*. Vol 7, 1942.

74. Shock, N. W. The effect of menarche on basal physiological functions in girls. *Am J Physiol*. 139: 288–2, 1943.

75. Stolz, H. R., and Stolz, L. M. *Somatic Development of Adolescent Boys*. New York: Macmillan & Co., 1951.

76. Addinsell, A. W. Premenstrual pregnancy in a girl aged 13 years. *Lancet*. 1: 791–2, 1905.

77. Allen, L. M. Pregnancy at the age of eleven years terminating in a natural delivery. *Maryland Med J*. 44: 416–21, 1901.

78. Bernheim, A. Diagnosis of early pregnancy. *Am Practitioner News*. (Louisville) 26: 330–2, 1898.

79. Beruti, J. A. *Archivo Clinica Obstetrica y Ginecologica* (Eliseo Canton). Vol. 2, 1943, pp. 227–31.

80. Boldt, H. J. Caesarian section on a child twelve years and eight months old with contracted pelvis, for severe puerperal eclampsia. *Postgraduate*. (New York) 20: 1251–67, 1905.

81. Bromberg, Y. M., and Brzenzinski, A. Labor in old and young primiparae. *J Obstet Gynaecol Br Empire*. 49: 672–81, 1942.

82. Buchanan, J. R. Precocious pregnancy. *Medical News*. (Philadelphia) 63: 15, 1893.

83. Carver, R. H. A case of early maternity. *Providence Med J*. 10: 83–5, 1909.

84. Chaschinsky, P. C. and Jerschow, S. I. Pubertas praecox and full term pregnancy in a girl 6 years old. *Zentralblatt für Gynäkologie*. 57: 2252–6, 1933.

85. Christian, W., and Rood, L. D. Precocious Pregnancy. *JAMA*. 31: 138, 1898.

86. Caso, R., and Vincent, D. P. Labor, puerperium and their complications in primiparas 12 to 16. *Revista Medico-quirurgica de Patologia Feminina*. 18: 402–10, 1941.

87. Comby, J. Maternites extraordinairement precoces. *Archives de Medicine des Enfants*. 41: 806–11, 1938.

88. Cook, W. W. Precocious pregnancies. *Chicago Med Obstetrician*. 1: 277, 1898.

89. Edmondson, R. H. An exceedingly early maternity. *Southern J Homeopathy*. 8: 116, 1890/91.

90. Fairfield, L. Mothers under 16. *Lancet*. 2: 61–3, 1940.

91. Findley, W. A precocious mother. *Medical News*. (New York) 70: 345, 1897.

92. Furtado, A. H. Precocious pregnancy, with report of a case. *Revista de Ginecologia e Obstetrica.* 2: 244–5, 1941.

93. Geets, V. Un cas de grossesse a 12½ ans. *Progress Medicale Belge.* 11: 153, 1910.

94. Gleaves, C. W. A young mother. *Medical Record.* (New York) 48: 715, 1895.

95. Godson, J. E. Pregnancy in the fourteenth year, previous to menstruation. *Br Med J.* 1: 743, 1893.

96. Gomez, F. L. Ten-year-old son. *Time.* 20 December 1954, p. 47.

97. Grebe, H. *Ueber Virginelle Konzeptionen.* Bonn, 1933.

98. Harris, W. J. Early pregnancy. *Br Med J.* 1: 179, 1890.

99. Hirts, B. C. Parturition in young girls. *University of Pennsylvania Medical Magazine.* (Philadelphia) 2: 601, 1889/90.

100. Hunter, Q. W. A review of the literature of precocious menstruation and precocious pregnancy. *Medical Age.* (Detroit) 24: 367–78, 1906.

101. Johnston, C. H. L. Precocious conception. *Lancet.* 2: 111, 1902.

102. Kerns, W. W. A very young mother. *Medical World.* (Philadelphia) 23: 26, 1905.

103. Livon, J. Grossesse et accouchement chez une femme de 12 ans et 5 mois. *Gazette de Gynécologie.* 21: 343–8, 1906.

104. Keane, H. L. A case of maternity at seven years of age. *Br Med J.* 2: 56, 1933.

105. Magruder, M. J. Report of a case of labor in a white girl of twelve years; postpartum hemorrhage; recovery. *Proceedings of the Orleans Parish Medical Society, 1899.* New Orleans, 1900, p. 23.

106. Marshall, G. B. Case of precocious pregnancy in a girl of 13 before the establishment of the menstrual function. *Transactions of the Glasgow Obstetrical and Gynaecological Society (1902–1904).* 4: 108, 1905.

107. Marx, S. Early pregnancy and labor. *Am Gynaecol Obstet J.* 4: 108, 1898.

108. Milner, J. Note on a case of precocious conception with subsequent delivery at full term. *Lancet.* 1: 1601, 1902.

109. Munde, P. F. Probable early pregnancy. *International Clinics.* (Philadelphia) 2: 273, 1898.

110. Nichols, T. B. Early maternity. *Medical Record.* (New York) 37: 273, 1898.

111. Penrose, L. S. Maternal age, order of birth and development abnormalities. *J Mental Sci.* 85: 1141–50, 1939.

112. Peo, E. Dystocia due to youth of mother. *Woman's Medical Journal.* (Toledo) 14: 49, 1904.

113. Pittman, V. I. Childbirth at the age of nine. *Am J Clin Med.* 15: 798, 1908.

114. Editorial. Quelques nouveau faits inedits de maternité precoce en France. *Semaine Medicale.* Paris 18: 119, 1898.

115. Reid, R. A. What is the maximum and what is the minimum age at which a woman is capable of bearing a child? *Mass Med J.* 10: 252–56, 1890.

116. Robertson, G. W. Pregnancy in childhood. *Br Med J.* 1: 142, 1902.

117. Russell, L. Precocious pregnancy and other unusual conditions. *Medical News.* (Philadelphia) 63: 219, 1893.

118. Shaffer, B. and Ricci, G. Age and primiparity; its influence; study based on 1,688 cases. *Boletin de la Sociedad de Obstetrica y Ginecologia de Buenos Aires.* 22: 47–66, 1943.

119. Scott, N. S. Precocious pregnancy. *JAMA.* 1: 1822, 1903.

120. Editorial. Sexual precocity. *Lancet.* 1: 1822, 1903.

121. Specht, A. *Ueber die Geburt beim Minderjährigen.* Kiel, 1917.

122. Stanley, D. F. Early motherhood. *Medical World.* (Philadelphia) 24: 346, 1906.

123. Tackett, J. R. A juvenile mother. *Mississippi Medical Journal.* (Meridian) 3: 248, 1893/94.

124. Thompson, R. R. A very young mother. *Medical Record.* (New York) 59: 423, 1901.

125. Underhill. What is the maximum and what the minimum age at which a woman is capable of bearing a child? *Transactions of the Obstetrical Society of Cincinnati.* 1: 179–82, 1877/79.

126. Vignes, H., and Hamelin, R. Les grossesse chez des jeunes enfants. *Presse Medicale.* 47: 1257–8, 1939.

127. Wales, E. B. Childbirth at twelve years of age. *Medical and Surgical Reporter.* (Philadelphia) 63: 129, 1890.

128. Warrington, L. P. Early pregnancy. *Medical News.* (Philadelphia) 61: 363, 1892.

129. Webster, J. C. Caesarian section on a girl thirteen years old. *Am J Obstet.* 43: 158–60, 1901.

130. Wilkinson, D. L. Precocious pregnancy; twin births. *JAMA.* 31: 424, 1898.

131. Willard, De F. Precocious pregnancy. *JAMA.* 31: 252, 1898.

132. Clark, J. et al. Pregnancy in the very young patient. *J Natl Med Assoc.* 54: 352, 1962.

133. Dodge, E. Discussion of paper by R. Aznar and E. Bennett. *Am J Obstet Gynecol.* 81: 940, 1958.

134. Escomel, E. La plus jeune mére du monde. *Presse médicale.* 47: 875, 1939.

135. Keane, H. A case of maternity at seven years of age. *Br Med J.* 2: 56, 1933.

136. Little mother. *Time.* 16 December 1957.

137. Reilly, W. A., Kisser, H., and Hinman, F. Pseudosexual precocity: the adrenal cortical syndrome in preadolescent girls. *Endocrinology.* 24: 91–114, 1939.

138. Jolly, H. *Sexual Precocity.* Springfield, Illinois: C. C Thomas, 1955.

139. Diamond, M., editor. *Perspectives in Reproduction and Sexual Behavior.* Bloomington: University of Indiana Press, 1968.

140. Huffman, J. W., editor. *The Gynecology of Childhood and Adolescence.* Philadelphia: W. B. Saunders, Co., 1968.

141. Gold, J. J. *Gynecologic Endocrinology.* 2nd edition. New York: Harper & Row, 1975.

142. Hartman, C. G. On the relative sterility of the adolescent organism. *Science.* 74: 226–7.

143. Smith, R. M., and Rubenstein, B. B. Adolescence of macaques. *Endocrinology.* 26: 667–9, 1940.

144. Quo, S-K. Mathematical analysis of the growth of man, with special reference to Formosans. *Hum Biol.* 25: 333–58, 1953.

145. Fried, R. I., and Smith, E. E. Postmenarcheal growth patterns. *J Pediatr.* 61: 362–5, 1962.

146. Schultz, A. H., and Snyder, F. F. Observations on reproduction in the chimpanzee. *Bull Johns Hopkins Hosp.* 57: 193–205, 1935.

147. Spence, K. W., and Yerkes, R. M. Weight, growth and age in chimpanzee. *Am J Phys Anthropol.* 22: 229–46, 1937.

148. Hartman, C. G. Studies in the reproduction of the monkey Macacus (Pithecus) rhesus, with special reference to menstruation and pregnancy. *Carnegie Institution of Washington, Contributions to Embryology.* 134: 23, 1932.

149. Mead, M. *Male and Female.* New York: William Morrow and Co., 1949, p. 202.

150. Benedek, T. *Psychosexual Functions in Women.* New York: The Ronald Press, 1952, p. 408.

151. Stieve, H. *Der Einfluss des Nervensystems auf Bau and Tätigkeit der Geschlechtsorgane des Menschen.* Stuttgart: Georg Thieme, 1951.

152. Selye, H. *Stress.* Montreal: Aota, Inc., 1950.

153. Wolff, H. G., editor. *Life Stress and Bodily Disease.* Baltimore: The Williams & Wilkins, 1950.

154. Wolff, H. G. *Stress and Disease.* 2nd edition. Springfield, Illinois: 1968.

155. Levi, L., editor. *Society, Stress and Disease.* London & New York: Oxford University Press, 1971/75/79.

7

THE PSYCHOLOGICAL AND PHYSIOLOGICAL IMMATURITY OF THE YOUNG MOTHER

In the preceding pages of this book we have seen that the evidence indicates that the adolescent female is biologically insufficiently developed to carry on the various functions of reproduction in a viably efficient manner. Prenatal accidents of various sorts and maternal and infant mortality rates are disproportionately high. Not only is the adolescent female physiologically unprepared, but in most cases she is behaviorally unready for the task of maternity. Adolescent female mice tend to kill or devour their newborn offspring more frequently than do mature female mice,* while adolescent chimpanzee mothers are generally quite incompetent to handle their newborn.[1-3] It would seem that behavioral mat-

*Interestingly enough, Little found that the size of litters produced by very young and by very old female mice is distinctly more variable than is the size of litters produced by females at the peak of their reproductive activity.

uration is quite as dependent upon physiological factors as is purely reproductive maturation. Here, too, there is evidence of some variation. Some very young mothers appear to handle their young well enough, but there can be little doubt that in general younger mothers fail to do so. Among human beings the example and instruction of older, more experienced females plays an important role in assisting the younger mothers to care properly for their children, but even so very young mothers are notoriously indifferent as compared with more mature mothers.

Evidence of the unreadiness of the adolescent to act as a responsible parent, is the finding of Dorothy Nortman that childhood mortality at ages one to four is 41 percent above average among children born to adolescent mothers.[4]

Mental development often appears to be seriously compromised in the children of adolescents. A Canadian study cited by Nortman found an increased prevalence of handicapped children of adolescent mothers, and a study of mental defectives found them over-represented among mothers under the age of 20 and over the age of 35. The J-shaped curve of risk stands out in high relief for young mothers, being lowest for women in their twenties, and higher for older mothers.

In the immature gravida iron stores, total skeletal stores, trace minerals, and vitamin A are substantially lower than in the adult female. From animal studies we know that inadequate vitamin A leads to increased teratogenesis. Giroud considers pregnancy in adolescence a contributing factor in the production of developmental defects of nutritional origin.[5,6]

A study, reported by Dr. Janet Hardy from the Johns Hopkins Child Development Study, on the development of 525 children born to girls who were 16 years or under at the time of delivery, found that at age four 11 percent of the children scored 70 or below on IQ tests compared with only 2.6 percent of the general population of four-year-olds. While in the general population approximately a quarter of four-year-olds will test out at 110 and above, only 5 percent of the children born of very young mothers tested that high.

School failure and behavior problems were also more prevalent in the study population. Dr. Hardy goes on to say, "Other serious problems such as child abuse, delinquent behavior and early pregnancies among the children themselves have been encountered. Yet some of the young mothers and their children have been successful. These mothers have completed their schooling, hold good jobs, have established satisfactory family life and their children are doing well and should be successful also."[7]

In short, as Dr. W. H. Baldwin says in her excellent survey of the subject, "To say that there are greater risks for children born to adolescent mothers does not mean, of course, that all such births will lead to problems any more than to say that delaying a birth until the twenties will guarantee that child a splendid life. We can say that the risk of negative outcomes for the pregnancy, the birth, or for the child's development are higher the younger the mother. Postponing a birth until a woman is in her twenties offers noticeably decreased risks in all respects." [8]

In nonhuman mammals it has been shown that prolactin plays a highly important role in producing maternal behavior in such animals, whether they have themselves recently given birth to young or not.[9] It is quite probable that the maternal inadequacies of the immature female are due at least in part to insufficient concentration of prolactin in her blood stream. This explanation would accord very well with all that we know concerning the physiological mechanism of adolescent sterility. However this may be, one of the best indications of the unpreparedness of the adolescent female organism for reproduction would appear to lie in the high mortality rate of infants born to young mothers, and also in the high maternal death rate of young mothers.[10,11]

But this is only a partial explanation, for there is good evidence that in all mammals maternal behavior is the result not exclusively of biochemical factors but of the reciprocal interaction between biochemical and social processes occurring within the mother.[12,13] As Brody put it, maternal behavior in animals "appears to involve physiological responses to instinctual organic needs—for nourishment, warmth, equilibrium, memory, and learning. And it appears to become explicable without reference to inborn adaptive mechanisms or to teleological purposes."[14] It is highly probable that physiological sexual maturation is correlated with emotional maturity and awareness of interpersonal relations, for these traits together with intellectual and emotional functioning have been found to be more highly developed in post- than in premenarcheal females.[15]

The secular trend toward earlier menarche (which appears to have ceased since the fourth decade of this century) doubtless contributes to the development of emotional, sexual, and heterosexual interests at an earlier age than was the case in previous generations. This is a possibility which has not yet received the attention it deserves.[16-18]

The relation between age at menarche and personality has been investigated by Shipman in 82 white American females with a mean age of 39 years. He found that the girls who reached menarche at 10 and 11 years, tended as adults to be more conservative and uncritical in their thinking, more lax and inexact, more trusting of others, and more group dependent. The middle menarche group, who attained menarche at age 12 and 13, tended to be more feminine, and to have a higher average number of children. The late menarche group, who attained menarche at 14 and older, were more dominant and aggressive, more given to critical thinking and ready to experiment, they were also more suspecting and self-controlled.[19]

The suggestion here is not that there is a genetic correlation between age at menarche and personality, but that social conditionings in the form of different responses, choices, and commitments during adolescence in relation to early or late menarche, may play a significant role in influencing personality development.

Buck and Stavraky found a positive and significant relationship between age at menarche and age at marriage for women in maternal age groups 25 to 19 and 30 to 34, but no such relationship in age group 35 to 39. The observations, made on 42,298 women, showed that there was a marked

Table 39
Frequency of Toxemias in Relation to Age

Age	No cases	No toxemias	Incidence Percent
12	3	0	0.0
13	19	8	42.1
14	77	27	35.0
15	163	30	18.4
16	372	60	16.1
Total	634	125	19.7

SOURCE: Marchetti and Menaker, *Pregnancy and the Adolescent,* 1950.

correlation between early menarche and early age at marriage.[20]

In addition, further evidence of the physiological unreadiness of the young mother for childbirth is the high frequency of toxemias and eclampsia. Marchetti and Menaker, in 634 young mothers ranging in age from 12 to 16 years inclusive (all blacks with the exception of seven whites) at the Gallinger Municipal Hospital of Washington, D.C., found the following frequency of toxemias in relation to age (Table 39). Of these females 93.7 percent were primagravidas.

From Table 39 it will be seen that the frequency of toxemia increases with the youthfulness of the mother. Comparing these frequencies with the incidence of toxemias in older mothers at the clinic, a great disparity is observed in favor the clinic group as compared with the young mothers (Table 40).

Table 40
Comparison of Incidence of Toxemias

Subjects	All Toxemias	Eclampsia
Clinic mothers	7.54	0.61
Young mothers	19.71	1.10

SOURCE: Marchetti and Menaker, *Pregnancy and the Adolescent*, 1950.

There were 2.5 percent more premature deliveries in the young than in the older group of mothers, and 3 percent as compared with 2.7 percent previable infants (those weighing 1 to 3 pounds). Clinic incidence of postpartum hemorrhage was 1.5 percent; in the young mothers it was 2 percent. Duration of labor was shorter in the young mother as compared with the clinic group mother.* Cesarian section less frequent, and the infant mortality rates were lower for the young than for the older clinic mothers, as shown in Table 40.[21]

Marchetti and Menaker admit that "From the point of view of frequency, as far as it can be determined, our incidence of 6.3 percent, or ratio of 1:17 is the highest thus far reported," and, indeed, it is a very high rate—ten times the rate for a similar population at the Harlem Hospital in New York, where it was 0.58 percent, 1:170 deliveries out of a total of 17,000 deliveries.[22] In view of such facts it is surprising that Marchetti and Menaker should conclude that "from the purely obstetrical point of view 16 years or less is the optimum age for the first baby"[23] (Table 41).

These authors are comparing a high infant mortality rate in the young mothers with an exceptionally high rate in the clinic mothers. Nevertheless, it is a remarkable fact that in their young series there were no maternal deaths. The

Table 41
Infant Mortality

Gestation	Total Deliveries	Deaths	Series Incidence	Clinic Rate
Term	540	4	0.6	1.7
Premature	78	8	1.3	2.6
Previable	19	12	1.9	2.0
Total	637	24	3.8	6.3

SOURCE: Marchetti and Menaker, *Pregnancy and the Adolescent*, 1950

*A doubtful advantage, since there is some evidence that a reasonably durable labor is helpful in readying the fetus for the assumption of his functions as a newborn. In this connection see Reference 21.

great improvement in antepartum and postpartum care during the preceding fifteen years must be held chiefly responsible for this fact. For the earlier years of this century the findings are quite different.

From Marchetti and Menaker's findings it would seem that some of the handicaps from which the young mother and her offspring suffer can be reduced by artificial means. It is, however, a serious question whether in the long run, from the purely biological viewpoint, more harm would not result than good from the encouragement of early maternity.

Seland in a study of delivery and its complications in a group of 397 primiparas between the ages of 14 and 17 years and in a control group of 1,070 married and 332 unmarried women aged 23 years who had been admitted to the larger maternity clinics in Oslo, Norway, found that more premature deliveries occurred in the 14–17-year-old group than in the control group. The average duration of a full-term pregnancy was nine months and 10 days. Eclampsia, average duration of labor, loss of more than 500 cc of blood, and infant mortality were greater in the adolescent than in the mature control group.[24]

Strandskov and Einhorn, analyzing the data taken from the United States Bureau of the Census yearly reports for the 15-year interval from 1922 to 1936 inclusive, involving mothers ranging in ages between 10 and 54 years inclusive, obtained the frequencies shown in Table 42.[25]

It will be seen from Table 42 that the mean percentage of stillbirths is highest, 6.70, in the youngest group of mothers (10–14), being higher than in mothers as old as 40 to 44 years, 6.01, but not as high in women at the close of the reproductive period in the age groups 45–49 (7.80) and 50–54 (8.09).

It is of interest to note the differences as between white and colored women. The mean stillbirth rates are uniformly higher in each age group in the colored than in the whites (Tables 43 and 44).

Strandskov and Einhorn consider that in addition to differences in competition between the growth processes of

Table 42
Means of 15 Yearly Percentages of Stillbirths (sex known) for 9 Different Age Groups of Mothers in Total U.S. Population, 1922–1936, Inclusive

Age Group	Total Live-Stillbirths Combined	No Stillbirths	Mean Percentage Stillbirth
10–14	37,663	2,502	6.70
15–19	3,639,829	142,958	3.93
20–24	9,177,407	280,643	3.06
25–29	8,171,776	245,241	3.00
30–34	5,615,423	197,260	3.51
35–39	3,516,069	162,211	4.61
40–44	119,511	70,756	6.01
45–49	117,072	9,161	7.80
50–54	2,408	197	8.09
Total	30,397,158	1,110,929	3.60

SOURCE: Strandskov and Einhorn, 1948.

Table 43
Means of 15 Yearly Percentages of Stillbirth (sex known and unknown) for 9 Different Age Groups of Mothers in White U.S. Population, 1922 to 1936 Inclusive

Age Group	Total Live-Stillbirths Combined	No Stillbirths	Mean Percentage Stillbirth
10–14	16,575	780	4.76
15–19	2,932,379	88,350	3.03
20–24	8,044,321	193,623	2.42
25–29	7,386,585	177,348	2.40
30–34	5,124,016	147,482	2.88
35–39	3,174,291	122,319	3.86
40–44	1,088,272	55,865	5.15
45–49	101,306	6,952	6.86
50–54	1,687	113	6.56
Total	27,869,432	792,832	2.82

SOURCE: Strandskov and Einhorn, 1948.

Table 44
Means of 15 Yearly Percentages of Stillbirth (sex known and unknown) for 9 Different Age Groups of Mothers in Black U.S. Population, 1922 to 1936, Inclusive

Age Group	Total Live-Stillbirths Combined	No Stillbirths	Mean Percentage Stillbirth
10–14	21,052	1,686	8.11
15–19	751,998	50,086	6.72
20–24	1,114,604	68,548	6.17
25–29	763,049	45,751	6.00
30–34	471,733	30,104	6.40
35–39	325,896	24,060	7.39
40–44	100,250	8,303	8.31
45–49	15,014	1,457	9.64
50–54	663	65	10.12
Total	3,564,258	230,060	6.51

SOURCE: Strandskov and Einhorn, 1948.

the mother and fetus, an even greater role in the causation of stillbirths in the young mother is to be ascribed to the inadequate development of the endocrine system, in particular the endocrines associated with implantation, maintenance of pregnancy, and parturition.

For a nonliterate society we have the observation of Williams on the Popoloca Indians of Puebla, Mexico, among whom the girls are said to reach puberty at 14 years on the average, at which time they are given in marriage. Early pregnancy is said to result in "quite frequent miscarriage as well as death of the mother at childbirth."[26]

High infant mortality rates are additionally characteristic of the reproductive history of the adolescent female.

It has been shown that some young mothers who have suffered repeated miscarriages in rapid succession have as their first full-term a Down's syndrome child.[27] The investigator, Dr. Lande-Champain, remarks that "the mongoloid pregnancy obviously marks the turning point from 'adolescent sterility'—documented through repeated miscarriages—to fertility."

This brings up the very interesting question of the correlation between maternal immaturity and the frequency with which immaturely or defectively developed infants are born to her as compared with the frequencies in mature mothers. The well-known high frequency of Down's syndrome births in older mothers[28-31] suggests that the inadequately developed reproductive system would be associated with a high frequency of inadequately developed offspring of such a system. There is a good deal of evidence in support of such an expected correlation.[29]

Engle's observations on mice in whom sexual maturity was precociously induced experimentally, that many eggs had not been fertilized or degenerated following fertilization, and the relatively high proportion of abnormal eggs recovered from immature heifers by Winters et al., suggests that the partial development of ova and embryos and their premature deaths are traits pecular to the immature females of all or most mammals.[32-33] Even in birds it has long been known that their first eggs are likely to be infertile. It is possible that some early uterine deaths in mammals may be due to the presence of lethal genes that prevent further development, but the high correlation between maternal age and infant mortality rate indicates that here, at least, other than genetic factors are operative.

The adolescent female is both psychologically and physiologically an immature organism who, in comparison with the female over nineteen and under thirty, is in many respects at a disadvantage. This holds true, in many cases not alone for herself but also for her offspring.

Battaglia, Frazier and Hellegers analyzed the outcome of 636 pregnancies occurring in girls aged 14 or under delivered at the Johns Hopkins Hospital between the years 1936/1960. The incidence of toxemia was 29.2 percent, contracted pelves 35.7 percent, prematurity 23.4 percent, immaturity 5.1 percent, and perinatal mortality 7.6 percent.[34]

Semmens and Lamers, in a study of 12,847 teenage pregnancies, among whom there were 2,955 multiparas, there was a total of 710 abortions or 5.5 percent. The incidence of abortion for a sample of 100 pregnancies under fifteen years of age was 2.1 percent. The low incidence of spontaneous abortion among teenagers, and especially in the youngest group, has been confirmed by many independent investigators.[35]

When one considers that after the age of thirty-five the incidence of abortion rises as high as 30 percent, the low incidence among teenagers is striking.[36]

Deaths associated with teenage pregnancy are also extremely rare. Semmens and Lamers in a series of 24,000 teenage pregnancies found not a single maternal death. Similarly, Israel and Woutersz reported no maternal deaths among the 3,995 teenage pregnancies they studied.[37]

In the under-15 age group Semmens and Lamers report a 3.2 percent perinatal mortality for mature infants, associated with prolonged labors up to 47 hours, as compared with a perinatal mortality in the 15 to 19 age group of 0.6 percent. The overall prematurity rate was 9.0 percent, as compared with the general population rate of 7.6 percent. In the under 15 age group preeclampsia with a weight gain of 26 to 35 pounds occurred in 4.2 percent; in the 15 to 19 age group the rate was 8.7 percent. The overall incidence of mild and severe preeclampsia in these 23,702 cases parallels the incidence of 6.0 percent reported by Israel and Woutersz. There were 7.4 percent labors prolonged beyond 21 hours in the mothers under 15 as compared to 3.8 percent in the 15 to 19-year-olds.

Preeclampsia, according to Israel and Woutersz, is three to four times more frequent in teenagers than the anticipated average for the clinic, held by some to be due to their careless dietary habits. But, as Israel and Woutersz say, there must be some other factor at work other than good prenatal care, since high rates of the condition occur among teenagers who enjoyed a satisfactory diet throughout pregnancy. Schmitz and Towne have reported that preeclampsia was the only significant complication among

200 unmarried teenage girls living in a Catholic charity home where they were well cared for and received well balanced meals.[38]

Since low birth weight is associated with a greater likelihood of mental retardation and other deficiencies as compared with birthweight within the normal range, this constitutes yet another group of factors unfavorable to teenage pregnancy. It is significantly more frequently associated with teenage maternity than it is at later ages in the twenties. In 1975, 13.0 percent of births to mothers 15 years of age weighed 2,500 grams or less, compared with 9.0 percent for births to 19-year-old mothers. The proportion of low-weight infants declined even further for births to mothers in the twenties: 7.1 percent for births to women aged 20 to 24 years, and 6.1 percent for births to women aged 25 to 29.[39]

Studies carried out by the National Research Council, and by the National Center for Health Statistics[41] show that the infant mortality rate is extremely high for very young mothers (less than fifteen years of age), and declines to a minimum in the early and late twenties, and then increases fairly sharply thereafter (Fig. 6). Neonatal mortality (less than 28 days after birth) is especially high when the mother is a teenager (Fig. 7).[40-41]

Prematurity rates are more than twice as high in teenagers under fifteen than at any other age, and significantly higher in teenagers at all ages than at any other age (Table 45).

Maternal mortality rates for white females is 15.1 percent in those under 20 and in white females aged 20–24 the rate is 11.0 percent.[42]

Maternal deaths from toxemia, excluding abortion, were 12.6 for nonwhites and 3.4 percent for whites; from abortion with sepsis 7.8 in nonwhites and 1.8 percent in whites, and from ectopic pregnancy 2.4 percent in nonwhites compared with 0.7 percent in whites. The rates for toxemia and abortion with sepsis about 50 percent higher for teenagers than for females 20-24 years of age in each ethnic group. Complications of pregnancy such as toxemia, pro-

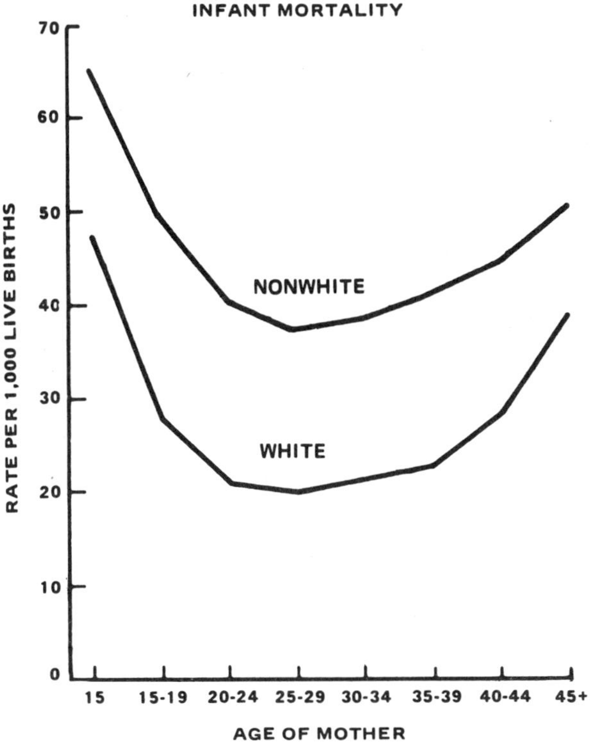

Figure 6. Infant mortality of white and nonwhite infants by age of mother: United States, 1960 Birth Cohort.

SOURCE: Working Group, "Relation of Nutrition to Pregnancy in Adolescence" in *Maternal Nutrition and the Course of Pregnancy*, Committee on Maternal Nutrition, Food and Nutrition Board, National Research Council (Washington: National Academy of Sciences, 1970), Table 3, p. 144.

longed labor, and iron-deficiency anemia, all occur more frequently in the teenage mother than in the mother of 20 and over.

A great deal of research has accumulated which shows that much of the poor obstetrical history of the teenage female is due to unfavorable socioeconomic conditions, and that when these conditions are improved there is a spec-

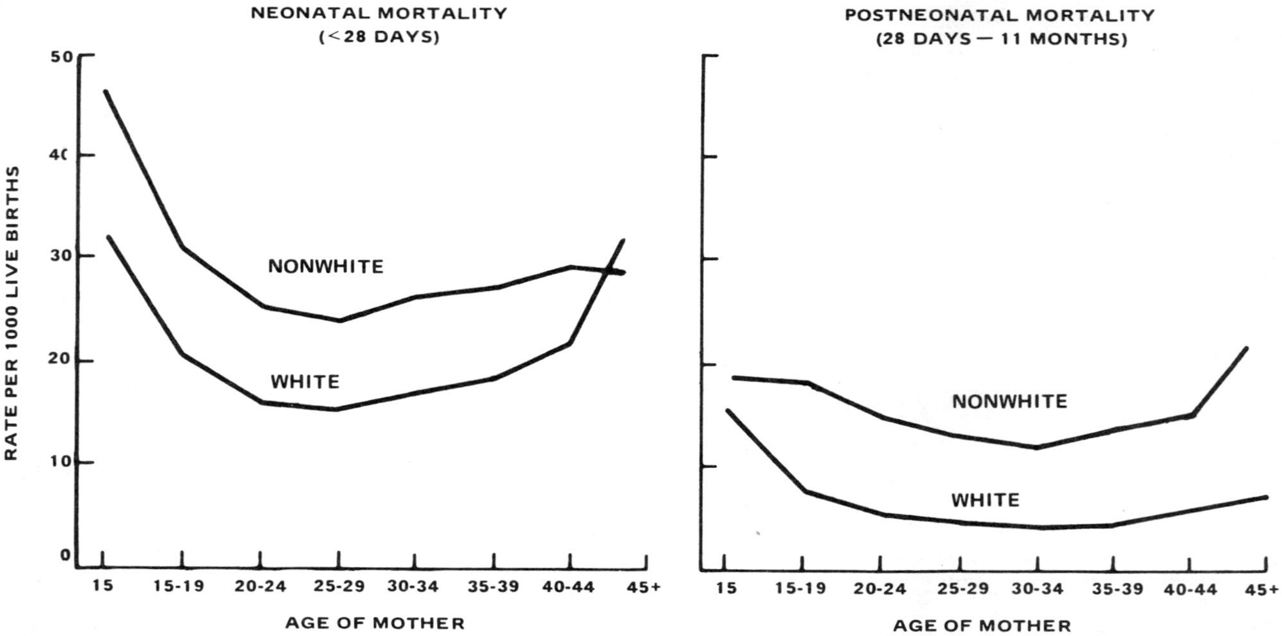

Figure 7. Neonatal and postneonatal mortality of white and nonwhite infants by age of mother: United States, 1960 Birth Cohort.

Table 45
Percent Low Birth Weight Infants by Age of Mother and Color: United States, January–March, 1950 and 1967

	Total			White			Nonwhite		
Age of mother	January–March 1950[a]	1967	Difference	January–March 1950[a]	1967	Difference	January–March 1950[a]	1967	Difference
Total	7.4	8.2	+0.8	7.0	7.1	+0.1	9.7	13.6	+3.9
Under 15 years	15.1	17.2	+2.1	15.9	12.5	−3.4	14.7	19.5	+4.8
15–19 years	9.0	10.5	+1.5	8.0	8.5	+0.5	12.0	15.7	+3.7
20–24 years	7.3	7.7	+0.4	6.9	6.7	−0.2	9.6	13.2	+3.6
25–29 years	6.7	7.2	+0.5	6.5	6.5	0	8.4	11.8	+3.4
30–34 years	7.2	7.9	+0.7	7.0	7.0	0	8.8	12.6	+3.8
35–39 years	7.7	9.1	+1.4	7.5	8.3	+0.8	9.0	13.3	+4.3
40–44 years	7.7	9.6	+1.9	7.5	9.1	+1.6	8.9	12.2	+3.3
45 years and over	6.1	8.6	+2.5	5.7	8.1	+2.4	7.4	10.8	+3.4

[a] Excludes all live births recorded in Massachusetts.

Source: Helen Chase, "Trends in 'Prematurity': United States, 1950–1967," *American Journal of Public Health*, 1970, Vol. 60, Table 8, p. 1978.

tacular improvement in all areas of the teenager's obstetrical performance.[43] Nevertheless, it also remains quite clear that a substantive contribution is made to the unfavorable obstetrical rates of the teenager by her physiological and emotional unreadiness for the tasks of childbearing and maternity.

REFERENCES

1. Little, C. C. *Genetics, Medicine, and Man*. Ithaca: Cornell University Press, 1947, p. 104.

2. Young, W. C., and Yerkes, R. M. Factors influencing the reproductive cycle in the chimpanzee; the period of adolescent sterility and related problems. *Endocrinology* 33: 121–54, 1943.

3. Yerkes, R. M. *Chimpanzees*. New Haven: Yale University Press, 1943, p. 68.

4. Nortman, D. Parental age as a factor in pregnancy outcome and child development. *Reports on Population Planning, No. 16*. New York: The Population Council, August 1974.

5. Giroud, A. *Nutrition of the Embryo*. Springfield, Ill.: Charles C Thomas Co., 1970.

6. Thomson, A. M. Pregnancy in adolescence. Edited by J. I. McKigney and H. N. Munro. In *Nutrient Requirements in Adolescence*. Cambridge, Mass.: MIT Press, 1976, p. 245–56.

7. U.S. Senate. School-Age Mother and Child Act, 1975. *Hearing Before the Sub-Committee on Health of the Committee on Labor and Public Welfare*. 94th Congress, 4 November 1975. U.S. Government Printing Office, 1976, p. 377.

8. Baldwin, W. H. Adolescent Pregnancy and Childbearing—Growing Concerns for Americans. *Population Reference Bureau Publication*. Vol. 31, no. 2, May 1977. Washington, D.C., p. 25.

9. Riddle, O., and Bates, R. W. The preparation, assay and actions of lactogenic hormone. Edited by E. Allen. In *Sex and the Internal Secretions*. Baltimore: The Williams & Wilkins Co., 1939, pp. 1088–1117.

10. Beach, F. A. *Hormones and Behavior*. New York: Paul B. Hoeber Inc., 1948.

11. Ford, C. S., and Beach, F. A. *Patterns of Sexual Behavior*. New York: Harper & Bros., 1953.

12. Freed, C. S., and Kroger, W. S. Psychologic manifestations of the menstrual cycle. *Psychosom Med.* 12: 229–35, 1950.

13. Rheingold, H. L., editor. *Maternal Behavior in Mammals*. New York: John Wiley and Sons, 1963.

14. Brody, S. *Patterns of Mothering*. New York: International Universities Press, 1956, p. 39.

15. Davidson, H. H., and Gottlieb, L. S. The emotional maturity of pre- and post-menarchial girls. *J Genet Psychol.* 86: 261–6, 1955.

16. Tanner, J. M. *Growth at Adolescence*. Springfield, Ill.: Charles C Thomas Co., 1955.

17. Zacharias, L., Wurtman, R. J., and Schatroff, M. Sexual maturation in contemporary American girls. *Am J Obstet Gynecol.* 108: 833–46, 1970.

18. Zacharias, L., and Lawther-Higgins, A. The age at menarche, at present and in the past. *Ms.* 1978.

19. Shipman, W. G. Age of menarche and adult personality. *Arch Gen Psychiatry.* 10: 155–9, 1964.

20. Buck, K., and Stavraky. The relationship between age at menarche and age at marriage among childbearing women. *Hum Biol.* 39: 93–102, 1967.

21. Montagu, A. *Touching: The Human Significance of the Skin.* 2nd edition. New York: Harper & Row, 1978.

22. Posner, C., and Pulver, M. An analysis of labor in young girls. *Am J Obstet Gynecol.* 30: 357–63, 1935.

23. Marchetti, A. A., and Menaker, J. S. Pregnancy and the adolescent. *Am J Obstet Gynecol.* 59: 1–8, 1950.

24. Seland, G. Delivery and its complications in adolescent primiparae. *Acta Obstet Gynecol Scand.* 35: 37–69, 1956.

25. Strandskov, H. S., and Einhorn, S. On the relation between age of mother and percentage of stillbirth in the total white and colored U.S. populations. *Am J Phys Anthropol.* 5: 187–98, 1948.

26. Williams, A. F. Notes on the Popoloca Indians of San Felipe, Otlaltepec, Puebla, Mexico. *Am Anthropol.* 48: 682–6, 1946.

27. Lande-Champain, L. The etiology of mongolism. *J Child Psych.* 3: 53–69, 1954.

28. Penrose, L. S. *The Biology of Mental Defect.* London: Sidgwick & Jackson, 1949.

29. Benda, C. E. *Mongolism and Cretinism.* New York: Grune & Stratton, 2nd ed., 1949.

30. Montagu, A. *Prenatal Influences.* Springfield, Ill.: Charles C Thomas Co., 1962.

31. Montagu, A. *Life Before Birth.* 2nd edition. New York: New American Library, 1978.

32. Engle, E. T. The pituitary-gonadal relationship and the problem of precocious sexual maturity. *Endocrinology.* 15: 405–20, 1931.

33. Winters, L. M., Green, W. W., Comstock, R. E. *Prenatal Development of the Bovine. University of Minnesota Agricultural Experiment Station, Technical Bulletin No. 151.* St. Paul, Minn.: 1942, p. 13.

34. Battaglia, F. C., Frazier, T. M., and Hellegers, A. E. Obstetric and pediatric complications of juvenile pregnancy. *Pediatrics.* 32: 902, 1963.

35. Semmens, J. P., and Lamers, W. M. *Teen-Age Pregnancy.* Springfield, Ill. Charles C Thomas Co., 1968, pp. 86–9.

36. Huber, C. P. Abortion. Edited by R. A. Kimbrough. In *Gynecology.* Philadelphia: J. B. Lippincott Co., 1965, pp. 471–4.

37. Israel, S. L., and Woutersz. Teenage obstetrics: cooperative study. *Am J Obstet Gynecol.* 85: 659–68, 1963.

38. Schmitz, H. E., and Towne, J. E. Adolescent primigravida. *Surgery, Gynecology and Obstetrics.* 84: 962–8, 1947.

39. Ventura, S. J. Teenage childbearing: United States, 1966–75. *Monthly Vital Statistics Report,* National Center for Health Statistics. Vol. 26 (suppl): 5, 1-15.

40. Working Group, Relation of nutrition to pregnancy in adolescence. In *Maternal Nutrition and the Course of Pregnancy.* Committee on Maternal Nutrition, Food and Nutrition Board, National Research Council, Washington: National Academy of Sciences, 1970.

41. Chase, H. Infant mortality and weight at birth: 1960 United States Birth Cohort. *Am J Public Health.* 59: 1618–28, 1969.

42. Mortality, Part A, Table 1-15. In *Vital Statistics of the United States,* Vol. II. Washington, D.C.: U.S. National Center for Health Statistics, 1967.

43. Menken, J. A. Teenage Childbearing: its medical aspects and implications for the United States Population. Commission on Population Growth and the American Future. Edited by Charles F. Westoff and Robert Parke, Jr. In *Research Reports, vol. 1, Demographic and Social Aspects of Population Growth.* Washington, D.C., 1972, pp. 333–53.

8

SUMMARY AND CONCLUSIONS

The evidence presented in the preceding chapters strongly supports the view that in the females of the mammals thus far investigated, mouse, rat, hamster, sheep, cow, macaque, chimpanzee, and human, the ability to reproduce is not in the large majority of cases synchronous with the appearance of the first estrus or the menarche. The physiological prerequisites of reproduction are ovulation *and* the maintenance of the developing fertilized egg within the uterus. Since these capacities do not generally develop until some time after the advent of the first menstruation, conception and reproduction are, therefore, in most cases, impossible during the menarcheal period. Before ovulation can take place the endocrine system and the soma must reach a certain level of development. When that level of development is attained ovulation occurs, and this is subsequently followed by the elaboration of prolactin (luteotrophin), and finally of progesterone. The organism is then said to be *nubile,* the phase of development during which it first becomes *viably* capable of conceiving and reproducing. Nubility and puberty are periods of development often confused with one another, but in reality they represent two very different developmental processes.

The interval from menarche (which is only one expression of puberty) to nubility is termed the *adolescent sterility period,* the period of adolescence during which the organism continues functionally to be reproductively sterile. Obviously, this period is not, because of its sterile character, overtly distinguishable from any of the preceding periods of development. It is here merely termed the adolescent sterility period in view of the necessity of emphasizing what appears to be the fact that menarche and the power to procreate are not synchronous events, but that the arrival of the one is usually separated from the development of the other by an appreciable interval of time during which the organism continues to be functionally sterile. This is not generally recognized to be the case. Menarche is most commonly believed to be the sign of the development of the capacity to reproduce, and the only interval at all recognized is that between menarche and maturity, the end of the period of adolescence.

Maturity is generally taken to be the period that commences at the time when the growth of the organism comes to an end, in humans at about 21 $\pm$ 2 years. This view, the evidence indicates, does not appear to be sound. The period of adolescence is arbitrarily taken to begin at menarche, but the power to procreate follows only at some time after this physiological function has developed. This is the period of ovarian maturation when ovulation becomes possible and when conception may follow, the period of nubility. This period, that is, nubility, is not coincident with maturity but represents the bridge, as it were, between puberty and maturity; maturity being attained only at an appreciable time after the development of nubility. The period of nubility is not by any means the best time for procreation, for it is a period still characterized by high maternal and infant mortality rates.[1-7] The best time for reproduction is unquestionably at the age of early maturity, when the female organism is fully prepared to undertake and carry through the processes of reproduction satisfactorily.

The physiological basis for the period of adolescent sterility is clear. It exists because the endocrine glands, which must pour their hormones into the blood stream in the necessary concentrations before ovulation and reproduction can occur, do not usually do so during the menarcheal or immediately postmenarcheal period, but only during a later phase of development, the period of nubility.

The period of nubility marks the progress of another series of steps in the maturation of the sexual equipment for the process of reproductive functioning. It does not by a long way represent the completion of that maturation, but only the beginning. It is necessary to distinguish quite clearly between the phases of puberty, nubility, and maturity.

Puberty may be defined as the period of the inception of those changes in the reproductive system that are exhibited in the gradual development of the secondary sex characters and the appearance of the first estrus or menarche. Physiologically this means that the pituitary gland now releases the follicle stimulating hormone (FSH) in increasingly greater concentrations, thus serving to produce secondary sexual differentiation. FSH causes rapid growth of the follicle, the central cavity of which fills with follicular fluid, which contains the hormones called estrogens. At this premenarcheal phase of development there is a sudden increase by some 40 percent in the FSH level. This high concentration of FSH now acts upon the genital tract in a specific manner and results in the appearance of the first estrus or menarche, and in the further development of the secondary sexual characters. During this and the immediately continuing phase of puberal development the organism is normally sterile.

To state this in somewhat greater detail:

The three hormones of the anterior pituitary involved in reproductive function, are (1) FSH which stimulates follicular growth to the preantral stage, (2) LH which initiates ovulation in the presence of prepared follicles, and (3) *prolactin,* which stimulates milk secretion by the breast, and the maintenance of the morphologic integrity and secretion of the corpora lutea. The two major glands of sex hormone secretion in the human female ovary are the theca interna of the follicle, which secretes estradiol, and the luteinized granulosa of the corpus luteum, which secretes progesterone.

The rising level of estradiol in the premenarcheal period initiates development of the secondary sexual characteristics, the uterus, vagina, pubic hair, and about a year before menarche pigmentation of the nipples and areolae, growth of the labia minora, and fat deposit in the labia majora. It is these changes, and not menarche, which really announce the advent of puberty—a term which in its very origin means the period of the appearance of hair on the pubes. A slight increase in the concentration of estradiol during this period triggers the release of gonadotropin releasing factor (GRF) from the hypothalamus, whereupon a surge of FSH from the anterior pituitary accompanied by a small amount of luteinizing hormone, together known as folliculotropin, induces growth of the follicle and the secretion by its theca interna of estrogen. With the development of a sufficiently high concentration of estrogen, probably also secreted by the granulosa cells of the follicle, a fall in estrogen level is followed by endometrial desquamation, a proliferative uterine endometrium is produced, and this results in the first menstruation menarche. Repetition of these events results in subsequent menstrual cycles. In the human female early menstrual cycles are especially liable to be anovulatory, but there is a great deal of variability. On average, in the human female the anovulatory cycles last about three years.

The changes we have described that are initiated during puberty, after the lapse of a variable period of time, are followed by a unique development, namely, ovulation. This marks the commencement of the period of nubility. Nubility may be defined as that period of development during which the functions of ovulation and reproductive potentiality are established in the organism.

The duration of the sterility period in mouse, monkey, chimpanzee, and human, would appear to be respectively one month, one to twelve months, four months to more than two years, and one month to seven years. Variability in respect of the duration of this period is, of course, the rule, but the precise limits of this variability in any of these groups remain to be determined. Nonetheless, it is clear that in the chimpanzee and in man this variability is appreciable.

Period		Changes
Adolescent sterility period 13–16 ± 1 year	Puberty	*FSH* increases by 40 percent Follicular growth First estrus or menarche Anovulatory cycles
Nubility 17–22 ± 2 years		*FSH* + *LH* Ovulation Corpus luteum Estrogen
Maturity 23–28 ± 2 years		*FSH* + *LH* Ovulation Estrogen *Prolactin* (*Luteotropin*) Corpus luteum Progesterone

THE ITALICIZED HORMONES ORIGINATE IN THE PITUITARY GLAND.

Figure 8

In about 25 percent of mice conception is possible at first estrus. In another study, it has been shown that in some macaques conception is possible at first estrus, and that depending upon the state of health of the animal the sterile period of the macaque may be but one month or it may be more than a year. In humans the observations of gynecologists and anthropologists would indicate that the interval between menarche and nubility is, on the whole, extremely variable.

The principal physiological changes characterizing the three main periods of reproductive development in the human female are briefly resumed in Figure 8.

Certainly, in the human species conception is possible at and even before menarche, and in many such cases the fetus may be carried uneventfully to term, but these are proportionately the exceptional cases. The adolescent sterility period is not one during which conception is impossible. Conception may occur, generally it does not. When it does occur during the adolescent sterility period the reproductive history of the individuals involved is generally marked by complications of pregnancy and a high infant mortality rate.

Physiologically the period of nubility is characterized by the following events.

The rising concentrations of estradiol in the post-menarcheal female trigger the luteinizing releasing hormone factor (LRF) which results in the release of luteinizing hormone from the hypothalamus. LH acts synergistically with FSH to cause rapid swelling of the follicle which then releases an ovum through the ovarian wall. The ovum is then taken up by the fimbria and passed into the uterine tube. The follicular cells that previously surrounded the ovum, become luteinized, that is, they become greatly enlarged and develop lipid inclusions that give the cells their distinctive yellow color, after which fact the secretory body they form in the wall of the ovary, the corpus luteum, is named. The corpus luteum produces most of the estrogen after ovulation under the stimulation mainly of LH.

Organization of the corpus luteum, and its maintenance as a temporary endocrine gland with its own secretions is not, however, achieved till the development of the third gonadotropic hormone, namely prolactin or luteotropin. It is the reorganizing action by prolactin of the corpus luteum that causes its granulosa cells to secrete progesterone. It is progesterone that prepares the uterus for pregnancy, and plays a role in preparing the breast (together with estrogens, corticosteroids, and growth hormone) for lactation.

It is not until this third stage, that secretion of prolactin by the anterior pituitary and the action of this upon the secretory functions of the corpus luteum to produce progesterone, that successful pregnancy becomes possible.

It is with the secretion of prolactin and progesterone that the period of reproductive maturity is initiated, a period which remains at its optimum for about five years. The evidence we have surveyed indicates that there exists an *absolute* period of adolescent sterility in most females during which reproduction cannot occur. This period, among females of any given population, is extremely variable in duration.

That a period of adolescent sterility normally exists in most human females is a fact substantiated as much by the observed phenomenon as by the nature of the physiological processes involved in the maturation of the reproductive system. This evidence strongly suggests that during early adolescence the organism must, in most cases, necessarily be sterile.

The fact that the existence of such a period of infertility in the postpuberal adolescent female has escaped the attention of most human beings is not as difficult to understand as may at first appear. Nor is it difficult to surmise how the belief came into being that the first menstruation represents a sign of the female's arrival at the procreative stage of development. What seems most probable is that it had been independently noted in various human groups that women remained sterile as long as they had not menstruated, that after the first menstruation women menstruated periodically, and that during their menstrual life alone were they capable of bearing children; for after the final cessation of these periods (menopause) they were again sterile. Women capable of menstruation were therefore considered to be capable of childbearing. Hence, the appearance of menstruation was taken to be the sign of the ability to procreate.

The inference is certainly a logical one, but it is not wholly in accordance with the facts, and unless the conditions are such as to make the observation a simple matter there is no possible way in which the falsifying factors entering into this inference would become apparent to sense. Since these conditions do not exist in most human societies it is practically universally believed that the first menstruation is a sign of the ability to procreate. It has been shown, in these pages, that there is good reason to believe that this is not the case.

The fact that in almost all nonliterate societies the female is married immediately after she attains menarche renders possible a sexual history for the person very different and much less trying than that which the person experiences in literate societies. In nonliterate societies premarital intercourse is usually, though not always, permitted. Since there is rarely any danger of pregnancy occurring before menarche the problem of illegitimate children does not arise. On the other hand, cultures in which marriage is postponed till some years after menarche, that is, until the postmenarchial nubile period, cannot permit premarital intercourse owing to the fact that the birth of many children out of wedlock would have a socially disrupting effect—for undoubtedly many of them would be born during the menarche-nubility interval. Hence, this additional impediment in such societies toward the recognition of an adolescent sterility interval.

We now come to the final practical conclusion of this study, namely, that the best time for conception, pregnancy, and childbirth in the human female is, on the average, at the age of 23 ± 2 years and for about five years thereafter. The society that would be healthy would do well to consider these facts, and would encourage the awareness of their significance for the female accordingly.

REFERENCES

1. Larini, O., and Giannizi, G. Le funzioni di maternita nelli giovani primiparae. Sorti di bambini nate da esse. *Ann Ostet Ginecol.* 53: 1177–234, 1931.

2. Hundh, G. On the problem of age and primiparity. *Acta Obstet Gynecol Scand.* 4(suppl): 137, 1926.

3. Semmens, J. B., and Lamers, W. M. *Teenage Pregnancy.* Springfield, Ill.: C. C Thomas Co., 1968.

4. Israel, S. L., and Woutersz, T. B. Teen-age obstetrics. *Am J Obstet Gynecol.* 85: 659–68, 1963.

5. Israel, S. L., and Deutschberger, J. Relation of the mother's age to obstetric performance. *Obstet Gynecol.* 24: 411–17, 1964.

6. Stearn, R. H. The adolescent primagravida. *Lancet.* 2: 1083–5, 1963.

7. Clark, J. F. et al. Pregnancy in the very young patient. *J Natl Med Assoc.* 54: 352–4, 1962.

APPENDIX A

THE RELATION OF AGE AT MENARCHE TO FERTILITY

It is sometimes inquired whether there exists any relation between early, "normal," and late incidence of menarche and fertility. The evidence bearing upon this question is, unfortunately, conflicting. In a study of the hospital records of 10,219 Edinburgh (Scotland) women, Kennedy found the mean age of menarche to be 15.03 ± 1.70 years. Kennedy states that "a correlation was not found between early or late incidence of puberty, and high or low fertility."[1]

Makuda and Hori investigated the relationship between menarche and sterility in 375 Japanese women from Kyoto, Japan. They found the mean age at menarche to be 14 years and eight months. They write that "the sterility rate is relatively low in the menarche of younger ages. But this rate is increased as the menarche is delayed. Especially a marked increase is proved in frequency in the sterile women who have later menarches than the age of 17 years."[2-3]

In Table 46, from Makuda and Hori, it would seem that there is a definite correlation between age at menarche and sterility rate in Japanese women.

Table 46
Age at Menarche and Sterility Rate

Age at Menarche	Sterility Rate*	Age at Menarche	Sterility Rate*
11–12	2.222	17–18	11.583
12–13	7.398	18–19	10.990
13–14	6.903	19–20	9.523
14–15	7.133	Above 20	20.000
15–16	7.495		

SOURCE: Makuda & Hori, Menarche and sterility, 1939.
* The percentage of the sterile patients to the sum of the sterile (*a*) and multiparae (*b*) whose menarches took place at the same age, is found and this is made the sterility rate ($\frac{a}{a+b} \times 100\%$ = sterility rate).

This correlation is supported for a Philadelphia group of 58 women reported by Mazer and Israel.[3] Of 58 patients who began to menstruate at the age of 17 or later, 21 (36 percent) had amenorrhea in later life and 8 (13 percent) were sterile without cause. In this connection it may be added that it is the experience of gynecologists that women with secondary amenorrhea often give a history of late menarche.

The suggestive findings of Makuda and Hori and of Mazer and Israel indicate the need for further research on larger series of women.

Treloar in a study of 324 University of Minnesota students, whose mean age at menarche was 13.6 years, found no relation between menarche, fertility, and menopause. Mean age at menopause was 49.5 years.[4]

REFERENCES

1. Kennedy, W. P. The menarche and menstrual type. *Obstet Gynaecol Br Empire*. 40: 792–804, 1933.

2. Makuda, T., and Hori, K., Menarche and sterility. *Japanese Journal of Obstetrics and Gynaecology*, 22: 190–4, 1939.

3. Mazer, C., and Israel, S. L., *Diagnosis and Treatment of Menstrual Disorders and Sterility*. New York: Paul B. Hoeber, Inc., 1951, p. 37.

4. Treloar, A. E. Menarche, menopause and intervening fecundability. *Hum Biol*. 46: 89–107, 1974.

APPENDIX B

ADOLESCENT STERILITY IN THE MALE

The existence of an adolescent sterility interval in the male has been suspected for some time, but proof of its existence became available only when it was put to experimental study by Webster and Young.[1-2] These investigators studied the sexual maturation of 20 male guinea pigs. The gross findings on each of the animals is set out in Table 46. The first intromission with pelvic thrusts occurred at 53 ± 0.8 days of age in 18 out of the 20 animals. In these animals the first ejaculation at 63.9 ± 3.0 days was sterile, and the first fertile ejaculation occurred at 81 ± 1.6 days of age. In only two animals, which matured late, did the first intromission and the first fertile ejaculation coincide. The interval between the first intromission and the first copulation constitutes the adolescent sterility interval in the male.

The failure of ejaculation accounted for most of the sterility during the early part of the period, and a deficiency

Table 47
Age of 20 Male Guinea Pigs at Sexual Maturation

Animal	Age in Days First Intromission	Age in Days First Ejaculation	Age in Days First Fertile Copulation
181	50	50	89
189	62	66	84
208L	52	59	74
210T	45	45	63
211L	40	56	83
214T	44	44	73
215T	96	96	96
231T	48	48	70
303	43	43	84
311	43	63	76
341	41	54	69
346	40	57	72
360	40	46	84
380L	55	93	93
456T	46	46	64
459L	55	55	90
461T	45	85	89
464T	75	89	89
465T	95	95	95
466T	52	52	85
Mean	53.5 0.8(P.E.)	63.9 3.0(P.E.)	81.1 1.6(P.E.)

SOURCE: Webster & Young: *Adolescent sterility in the male guinea pig,* 1951.

of spermatozoa in the ejaculate appears to have been responsible for the sterility later, during the period of sterile ejaculations. Webster and Young found that in the guinea pig as in the rat, the development of spermatogenesis and maturation of the sex drive are not necessarily strictly correlated in time.

These findings are paralleled in man. The prepubertal male develops erections long before he is capable of ejaculations, and his first ejaculations are said to be low in spermatozoa.[3]

It would appear to be a fairly safe inference that the first ejaculation in boys corresponds to the menarche in girls. Ramsey found that in 286 white boys from a Middle Western city of over 100,000 population, and in 37 black boys from an eastern residential town of approximately 8,000 population, the median age for first ejaculation was the same in both groups, 13.8 years. The median age for voice change was 13.7 for black males and 13.4 years for white males. The median age for first appearance of pubic hair for the black group was 13.3 years, and for the white group 13.6 years.[4]

These findings are virtually identical with those obtained by Kinsey.[5] In response to a query from Webster and Young, Kinsey stated that "Unpublished data gathered by the group studying sex behavior at Indiana University indicate that there may be a period of some years of adolescent sterility in the human male. The data confirm the previously published work on adolescent sterility in the human female and indicate something of the same sort for the human male. We have a very few records of younger adolescent males being responsible for pregnancies even in groups where coitus is frequent and no form of contraception is employed. Some data exist on the presence of sperm in the first ejaculates of human males, but we are inclined to believe that sperm counts may be too low in the younger human male to effect fertilization. We would emphasize that there must be considerable individual variation on all these matters."

REFERENCES

1. Lataste, F. Des charactères de l'aptitude du dobaye male a la reproduction. *Compte Rendus du Societé de Biologie*. Series 9 4: 674–6, 1892.

2. Webster, R. C., and Young, W. C. Adolescent sterility in the male guinea pig. *Fertil Steril*. 2: 175–81, 1951.

3. Baldwin, B. T. The determination of sex maturation in boys by a laboratory method. *J Comp Psychol*. 8: 39–43, 1928.

4. Ramsey, G. V. Sexual growth of Negro and white boys. *Hum Biol*. 22: 146–9, 1950.

5. Kinsey, A. C. et al. *Sexual Behavior in the Human Male*. Philadelphia: W. B. Saunders Co., 1948.

APPENDIX C

AGE AT MENARCHE AND DURATION OF ADOLESCENT STERILITY

Is there a relationship between age at menarche and the duration of the adolescent sterility period? There is some evidence to suggest that such a relationship does exist. Early age at menarche appears to be related to an absolutely longer sterility period, that is to say, the earlier the age at menarche the longer is the duration of the adolescent sterility period, whereas at later menarcheal ages the duration of the adolescent sterility period is absolutely shorter. This is graphically shown in Figure 9, which represents Shuttleworth's analysis of the data presented by Mills and Ogle.[1,2] It will be seen from this graph that girls who menstruate first at 11.5 years are, on the average, likely to exhibit an adolescent sterility period lasting somewhat more than five years, whereas girls who first menstruate at 16.5 years are, on the average, likely to be characterized by an adolescent sterility period of about 1.75 years.

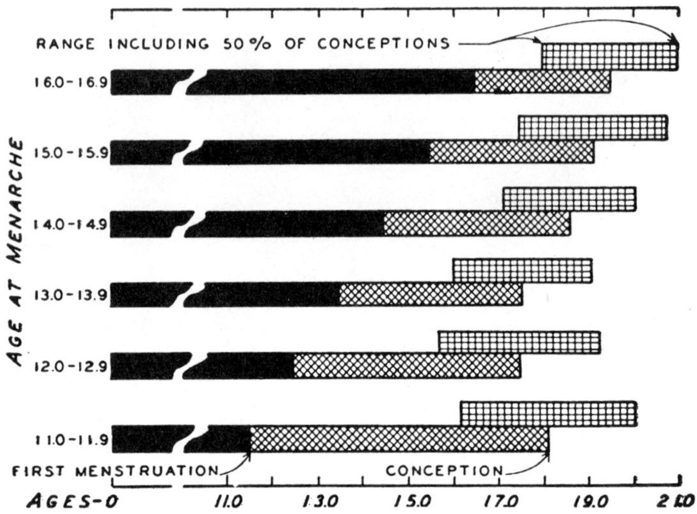

Figure 9. Average age and range of ages at conception leading to first delivery of an illegitimate child according to age at first menstruation, data for whites and Negroes combined. Note the lag between age at first menstruation and age at conception among all groups suggesting a period of adolescent sterility. The lag is considerably greater among girls who menstruate early. Based on data of Mills and Ogle.

The observations of most of the investigators cited in the present volume support this conclusion.[3,4] The obvious inference is that while age at menarche may be very variable, the age at which the female is ready for conception is much less so. This means that the duration of the adolescent sterility period is variable in any given population, but that however early or late menarche appears an appreciable amount of time must, on the average, elapse before the female is ready for conception. In early menarcheal females this will usually occur at an earlier age than in late menarcheal females, even though a greater absolute period of time elapses in the former than in the latter before conceptional capacity is attained.

REFERENCES

1. Shuttleworth, F. K. The adolescent period: a graphic analysis. *Monographs of the Society for Research in Child Development*. Vol 14, No. 1, Figure 126, 1949.

2. Mills, C. A. and Ogle, C. Physiologic sterility of adolescence. *Human Biology*. 8: 607–15, 1936.

3. Kiernan, K. E. Age at puberty in relation to age at marriage and parenthood: a national longitudinal study. *Ann Hum Biol*. 4: 301–8, 1977.

4. Douglas, J. W. M. The age of reaching puberty: some associated factors and some educational implications. The *Scientific Basis of Medicine Annual Reviews*. 91–105, 1966.

APPENDIX D

ADVANCED AGE AT CHILDBIRTH

The oldest recorded and authenticated age at which a woman has had a baby is 57 years four months and seven days. This lady, twelve years after her second marriage on 10 April 1944, on 18 October 1956 gave birth to her firstborn child, a daughter.[1] However, such late maternity is extremely rare, and unless the evidence is carefully checked most reported births to mothers of advanced age are to be viewed with some reservation. The National Center For Health Statistics at Washington checking such reports for mothers alleged to be 50 or over found that "Of the 55 births originally counted, there is no absolutely certain evidence that a single one occurred to a mother who was even 50 years of age."[2]

REFERENCES

1. Flack, H. Having a baby at fifty-seven. *Family Doctor,* 15: 578–9, 1965.

2. The age-of mother category—50 years and over. *Monthly Supplement, Vital Statistics Report*. Vol. 14, no. 9. Washington, D.C.: National Center for Health Statistics, November 1965.

INDEX

Abelam tribe of New Guinea, 15
adolescence, 3, 4, 54
  biology of, 4
adolescents
  as mothers, vii, 192
  growth spurt, 167, 175
  pregnancy in, vii, 200
    Caesarian section, 196
    disadvantages of, vii
    Down's syndrome child, 199
    duration of labor, 196
    infant mortality, 196, 201
    maternal mortality, 202
    miscarriage, 199
    premature delivery, 201–202
    stillbirths, 197–199
    undesirability of, viii, 161
adolescent sterility
  in human females, 54, 83–148, 208, 214–217
    anovulatory cycles, 165, 172, 173
    biology of, 161–182
    duration, 225–227
  in human males, 221–223
advanced age at childbirth, 229
Angami Nagas of Assam, 7
Apa Tanis of Eastern Himalayas, 20
Australian aborigines
  Adnjamatana tribe, 16
  Cape York, 17, 18
  Central Australia Walbiri, 18
  Simpson's Bay, 17
  Ooldea region, 17
  Port Lincoln, 18, 97
  Western South Australia, 17
  Yirrkalla, 17
Azande of the Sudan, 24

Baiga tribe of India, 14
Basuto tribe of South Africa, 99
Bengalee women, 23
Busama of North-Eastern New Guinea, 14
child marriage, 40, 43 (*see also under* marriage)
chimpanzee, 61–78, 191
  age of menarche, 64, 68, 69
  maternal inadequacies, 62, 63, 64, 71, 78, 191
  sexual skin, 61, 63, 64, 70, 72, 76, 77
  sterility interval, 62, 66, 68, 77
Chinese women, 103, 104
  Cochin, 99, 100, 102
  Hong Kong, 105, 106, 107
  South Fukien, 102
coitus interruptus, 9, 14
conception
  optimum age for, 214
  precocious, 136–147
contraception, 23–25
corpus luteum, 40, 43, 213, 214
cow, 49, 200

D'Entrecasteaux Islanders, 15
Dobuans of the Western Pacific, 9, 15
Donaldson's law of equivalent ages, 53
Dubum Dani of West New Guinea, 22

Eastern Rengmas of Assam, 14
eclampsia, 195
Eddystone Islanders of Melanesia, 7, 23
Eskimos
  Baffin Island, 21
  Nunamiut, 21
estradiol, 76, 171
estrogen, 58, 174, 212, 213

fecundity, 25, 85, 90, 91, 92
fertility, 25

fertilization, 55
Fijians of Vanua Levu Quain, 19
follicle stimulating hormone (FSH), 58, 170, 171, 173, 211, 212, 213
Fuegians of South America, 97

golden hamster, 47
gonadotropic hormone, 58, 164, 214
Gonds of India, 19
gorilla *(Gorilla gorilla berengei),* 73
(Gorilla gorilla), 74–75

Hindu women, 99, 100, 109–117
human chorionic gonadotropin, 173
hypothalamus, 171

Ifaluks of the Carolines, 21
Ifuago of the Philippines, 14
immaturity, 85, 161
implantation, 41, 55
impregnation, 172
infecundity, 25
infertility, 25
interstitial cell stimulating hormone (ICSH), 166

Java macaque *(Macacus cynomolgus),* 57–59
Jivaro of Ecuadoran Amazon, 22

Kaska Indians, 20
¡ Kung Bushman of Botswana, 22

Lepcha women, 129–132
luteinizing hormone (LH), 57, 58, 166, 171, 173 211, 213, 214
luteinizing releasing hormone (LRH), 171, 214
Luzon, Philippines, 16

marriage, 2, 126, 129
  and duration of sterility, 225–227
  child, 2
  -conception interval, 147–148
  minimum age for, 2, 3
maturity, 4, 54, 210
menarche, vii, 4, 5, 94, 95, 211, 215
  age at 11, 130–135, 194
    in various populations, 108, 118–124, 125
  and climate, 11, 12
  and intelligence, 11, 12
  genetic influences on, 11
  nutritional factors, 11, 13
  socioeconomic factors, 11–13
  weight at, 54, 167–170, 179
menstruation (*see also under* menarche)
  -conception intervals, 127, 143, 145, 146, 147, 219–220
motherhood (*see also under* adolescents as mothers)
  physiologic capacity, 3, 191–203
  psychological capacity, 3, 191–203
mouse, 34–46, 191
  adolescent sterility, 46
  corpora lutea, 40, 41, 42, 214
  estrus, 34, 40
  factors of sterility, 36, 38
  ovulation, 40, 41, 43
  puberty, 34, 38, 42
  vaginal plate, 39
  vaginal plug, 39
Muria of India, 19, 24, 130, 132–135, 181

natality, 25
North-West Malekulese of the New Hebrides, 6, 19
nubility, 59, 86, 87, 101, 208

ovarian activity, 5
ovulation, 55, 172, 212

penguin, Yellow-eyed, 31–34
pregnancy, adolescent (*see under* adolescents)
prolactin, 210, 211
puberty, vi, 4, 54, 87, 101, 129, 170, 175, 210, 211, 213
Pygmies of Ituri Forest, 21

reproductive system
  capacity, 5, 162, 211
  cultural factors, 181–182
  physiology, ix, 163
    cycles, 165, 166, 169
    fertilization of ovum, 162
    maturity of ovum, 162, 165
    optimal conditions, 162
    weight factor, 168
rhesus monkey *(Macaca mulatta),* 53–57
  adolescent sterility, 54, 55, 56
  body weight, 57
  sexual skin, 56
Romonum (Truk) Micronesia, 19

secondary sexual characters, 171
sheep, 47
Southern Massim of New Guinea, 7
sterility, 25, 93
  adolescent (*see under* adolescent sterility)

Tahitians, 22, 23
toxemia, 195, 200, 202
Trobriand Islanders of North-Western Melanesia, 8
tropics, 11

Wogeo Islanders of New Guinea, 10, 23, 105